Clearer Picture of the Spatiotemporal Growth of a Pull-Apart Basin - High-Resolution Geophysical Study at the Termination of an Arc-Bisecting Fault, Southwest Japan

Authored by Yasuto Itoh et al.

Published in London, United Kingdom

Clearer Picture of the Spatiotemporal Growth of a Pull-Apart Basin - High-Resolution Geophysical Study at the Termination of an Arc-Bisecting Fault, Southwest Japan

http://dx.doi.org/10.5772/intechopen.110456

Authored by Yasuto Itoh et al.

Contributors

Yasuto Itoh, Shigekazu Kusumoto, Yousuke Teranishi, Takeshi Kozawa, Hitoshi Tsukahara, Fumitoshi Murakami, Akio Hara, Taiki Sawada

First published in London, United Kingdom, 2023 by IntechOpen

IntechOpen is the global imprint of INTECHOPEN LIMITED, registered in England and Wales, registration number: 11086078, 5 Princes Gate Court, London, SW7 2QJ, United Kingdom

British Library Cataloguing-in-Publication Data

A catalogue record for this book is available from the British Library

Additional hard and PDF copies can be obtained from orders@intechopen.com

Clearer Picture of the Spatiotemporal Growth of a Pull-Apart Basin - High-Resolution Geophysical Study at the Termination of an Arc-Bisecting Fault, Southwest Japan

Authored by Yasuto Itoh et al.

p. cm.

Print ISBN 978-0-85014-369-0

Online ISBN 978-0-85014-371-3

eBook (PDF) ISBN 978-0-85014-370-6

For EU product safety concerns:
IN TECH d.o.o., Prolaz Marije Krucifikse Kozulić 3, 51000 Rijeka, Croatia,
info@intechopen.com or visit our website at intechopen.com.

Meet the Main Author

Yasuto Itoh received his Ph.D. at Kyoto University, Japan. He is currently a professor at the Graduate School of Sustainable System Sciences of Osaka Metropolitan University, Japan. He conducts research into tectonics, stratigraphy, structural geology, and paleo-/rock magnetism. He has published more than 150 papers on the deformation and amalgamation process of active plate margins, seismic hazard assessment of active faults, the paleoenvironment of the East Asia region, and the mechanism of sedimentary basin formation/development.

Co-Authors

- Shigekazu Kusumoto
  Institute for Geothermal Sciences, Kyoto University, Oita, Japan
- Yousuke Teranishi
  JGI, Inc., Tokyo, Japan
- Takeshi Kozawa
  JGI, Inc., Tokyo, Japan
- Hitoshi Tsukahara
  JGI, Inc., Tokyo, Japan
- Fumitoshi Murakami
  JGI, Inc., Tokyo, Japan
- Akio Hara
  JGI, Inc., Tokyo, Japan
- Taiki Sawada
  JGI, Inc., Tokyo, Japan

Contents

Preface

As the most vigorous material recycling sites on Earth, convergent plate margins have been studied from the viewpoints of tectonics, sedimentology, and global environmental changes. They are, however, often characterized by inaccessible topography and a high geothermal gradient, which hinder surface and subsurface investigations. This book presents the outcomes of the latest case study specifically focused on an active pull-apart basin along the northwestern Pacific margin.

Chapter 1 is the prologue that introduces us to the tectonic context of the study area.

Chapter 2 provides the geological framework of the conspicuous depression generated at an arc–arc junction.

Chapter 3 gives an overview of the tectonic sag based on gravitational analyses and reproduces its evolution by means of numerical modeling approaches.

In Chapter 4, fine sub-seafloor structures within Beppu Bay, a part of the study area, are revealed by a newly developed technique of 3D high-resolution seismic surveying. The chapter also highlights the principal data acquisition/processing sequence and the obtained results.

Based on this geophysical accomplishment, we discuss the spatiotemporal deformation processes in the Beppu Bay basin at the termination of an active transcurrent fault and reconstruct paleoenvironments with the aid of seismic attribute analysis in Chapter 5.

Finally, in Chapter 6 we summarize all the scientific findings. Our multidisciplinary research sheds light on the understanding of the complex architecture of convergent margins.

Yasuto Itoh
Graduate School of Sustainable System Sciences,
Osaka Metropolitan University,
Osaka, Japan

Chapter 1

Prologue: Tectonic Context of the Study Area

Yasuto Itoh

Pull-apart basins are formed at releasing morphologies of strike-slip faults. They have an important role in transferring mass at oblique convergent margins, at which strong simple shear provoked by a subducting slab eventually generates an arc-bisecting strike-slip crustal break [1]. Due to the terminal propagation of a group of sub-parallel transcurrent faults, the depocenter of a pull-apart basin migrates along the forearc and is progressively filled by terrigenous sediments [2]. The EW-trending Median Tectonic Line (MTL) active fault system in southwest Japan (**Figure 1**) is a

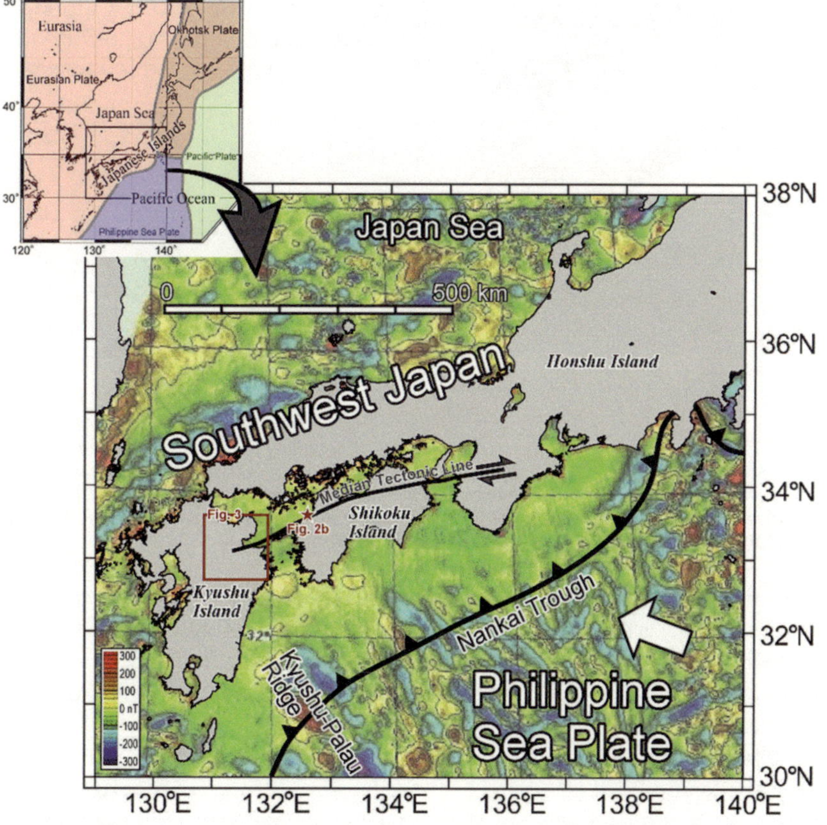

Figure 1.
Index maps for the northwestern Pacific margin and Southwest Japan after [3]. The open arrow denotes the late Quaternary relative motion of the Philippine Sea plate [4].

IntechOpen

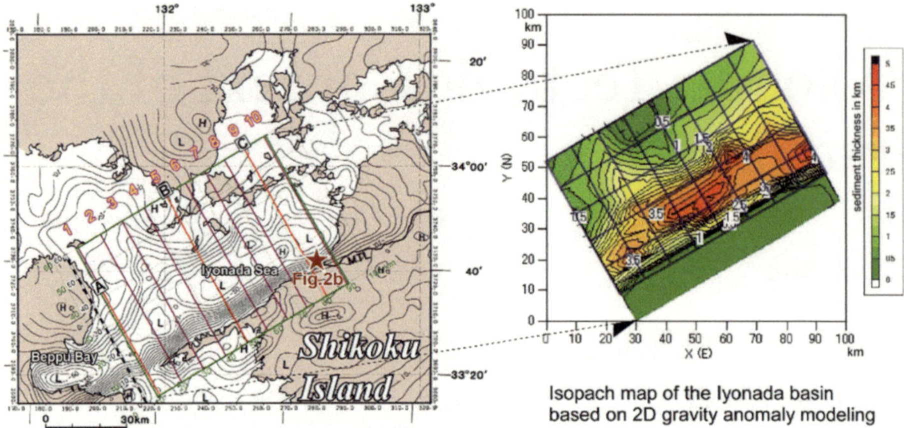

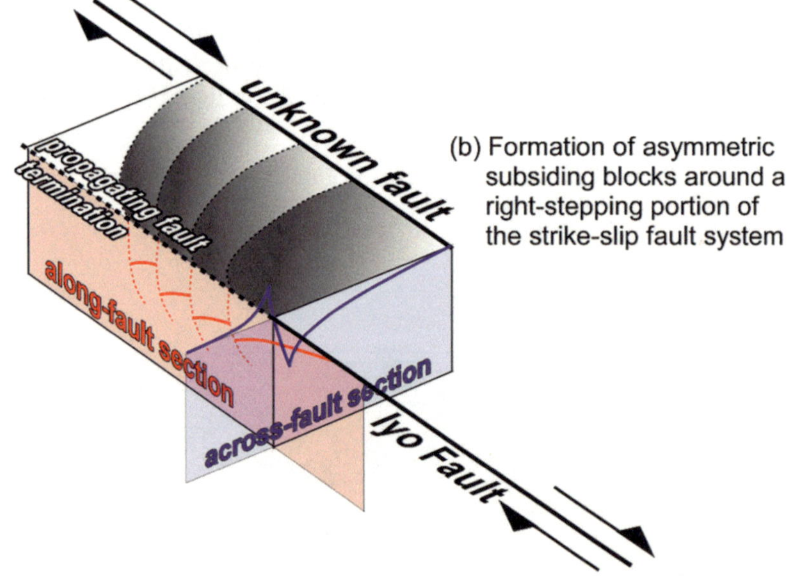

Figure 2.
(a) Results of volumetric study of the Quaternary basin around the northwestern coast of Shikoku Island [5]. Contours in the index map (left) are Bouguer anomalies at 5-mGal intervals, and numbered lines are for 2D gravity anomaly modeling. (b) Hypothetical process of pull-apart basin formation on the northwestern coast of the Shikoku Island [3, 6]. See **Figures 1** *and* **2a** *(left) for the location.*

typical arc-bisecting fault driven by oblique subduction of the Philippine Sea Plate through the late Quaternary [4]. The western part of the regional dextral fault is accompanied by releasing bends and steps, and previous studies [3, 5, 6] revealed the dimensions and buildup sequence of a pull-apart sag near northwest Shikoku Island (**Figures 1** and **2**).

In this book, we concentrate on the Beppu Bay basin in the central part of Kyushu Island (**Figure 2a** and **3**). It is located near the western termination of the MTL active fault system and occupies the eastern part of the volcano-tectonic depression of the Hohi Volcanic Zone [8], as shown in **Figure 3**. Although the basinal area has an absence of deep boreholes that could constrain the stratigraphic framework, its

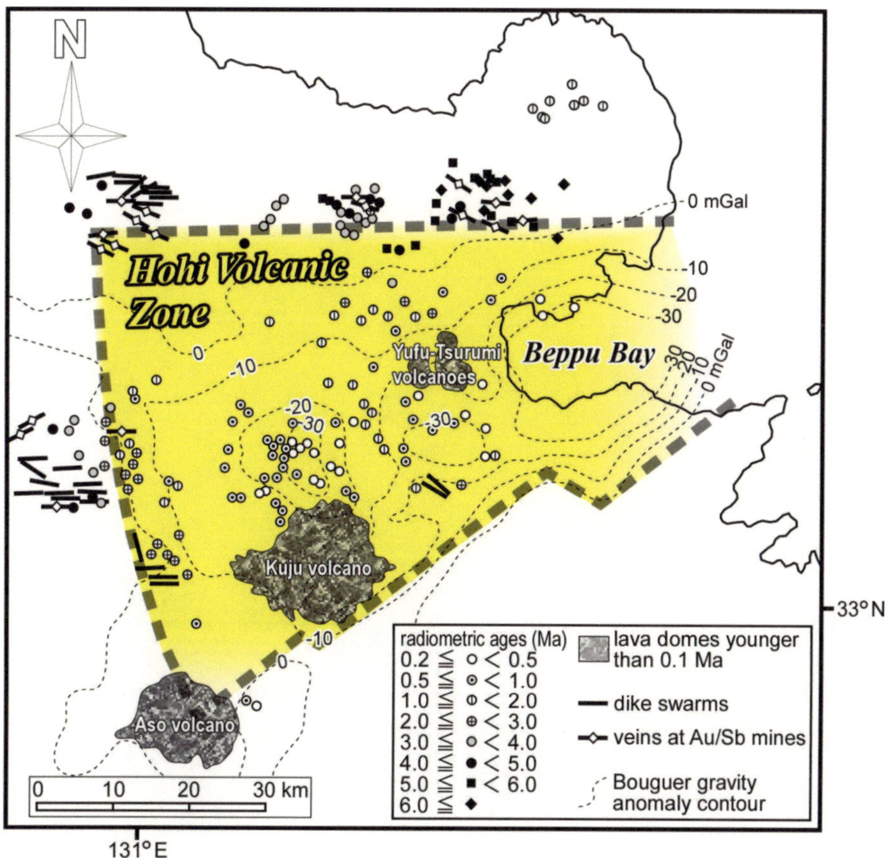

Figure 3.
Extent, geologic features, and gravity anomaly around the Hohi volcanic zone modified from [7]. See **Figure 1** *for the mapped area.*

structural outline and evolutionary process have been addressed based on gravimetric analysis [9] and numerical modeling [10], which has an extraordinary advantage to figure out dimensions and structural buildup within a volcanic basin. Deep architecture of the basin shown by modeling studies has been confirmed by reflection seismic interpretations [7, 8, 10]. The previous works have suggested that the architectural characteristics of the basin are governed by a temporal transition of the active segments of the MTL. As it is an interesting case of a tectonically controlled estuary system, the authors introduced the latest technology of 3D high-resolution seismic (3D-HRS) surveying. 3D-HRS has the ability to visualize exquisite shallow fractures and sedimentary facies that have direct linkage with the late Pleistocene MTL transition and eustatic sea-level changes that are recorded around the bay, respectively. After geological and geophysical reviews in Chapters 1 and 2, the methodology and effect of the 3D-HRS is fully described in Chapter 3. Then, we demonstrate the notable stress gradient around the pull-apart basin, which provokes the Pleistocene depocenter migration and delineate the subsurface paleo-drainage system by means of a conventional 2D seismic survey [7] and the seismic attributes of the 3D-HRS dataset, respectively, in Chapter 4. Finally, our achievement and perspective for future work is summarized in Epilogue.

Author details

Yasuto Itoh
Osaka Metropolitan University, Osaka, Japan

*Address all correspondence to: yasutokov@yahoo.co.jp

References

[1] Fitch TJ. Plate convergence, transcurrent faults, and internal deformation adjacent to Southeast Asia and the western Pacific. Journal of Geophysical Research. 1972;**77**:4432-4460

[2] Noda A. Strike-slip basin - its configuration and sedimentary facies. In: Itoh Y, editor. Mechanism of Sedimentary Basin Formation - Multidisciplinary Approach on Active Plate Margins. Rijeka: InTechOpen; 2013. DOI: 10.5772/56593

[3] Itoh Y. Time-series analysis of crustal deformation on longstanding transcurrent fault - structural diversity along Median Tectonic Line, Southwest Japan, and tectonic implications. In: Cengiz M, Karabulut S, editors. Earth's Crust and its Evolution - from Pangea to the Present Continents. London: IntechOpen; 2022. DOI: 10.5772/intechopen.101329

[4] Nakamura K, Renard V, Angelier J, Azema J, Bourgois J, Deplus C, et al. Oblique and near collision subduction, Sagami and Suruga troughs - preliminary results of the French-Japanese 1984 Kaiko cruise, leg 2. Earth and Planetary Science Letters. 1987;**83**:229-242

[5] Itoh Y, Kusumoto S, Takemura K. Characteristic basin formation at terminations of a large transcurrent fault - basin configuration based on gravity and geomagnetic data. In: Itoh Y, editor. Mechanism of Sedimentary Basin Formation - Multidisciplinary Approach on Active Plate Margins. Rijeka: InTechOpen; 2013. DOI: 10.5772/56702

[6] Itoh Y. Gunchu Formation - an Indicator of Active Tectonics on an Oblique Convergent Margin. Germany: LAP LAMBERT Academic Publishing; 2015. p. 76

[7] Itoh Y, Inaba M. Fault-Controlled Processes of Basin Evolution: A Case on a Longstanding Tectonic Line. New York: Nova Science Publishers, Inc.; 2019. p. 81

[8] Itoh Y, Takemura K, Kamata H. History of basin formation and tectonic evolution at the termination of a large transcurrent fault system: Deformation mode of Central Kyushu, Japan. Tectonophysics. 1998;**284**:135-150

[9] Kusumoto S, Fukuda Y, Takemoto S, Yusa Y. Three-dimensional subsurface structure in the eastern part of the Beppu-Shimabara Graben Kyushu, Japan, as revealed by gravimetric data. Journal of the Geodetic Society of Japan. 1996;**42**:167-181

[10] Itoh Y, Kusumoto S, Takemura K. Evolutionary process of Beppu Bay in Central Kyushu, Japan: A quantitative study of the basin-forming process controlled by plate convergence modes. Earth, Planets and Space. 2014;**66**:74. Available from: http://www.earth-planets-space.com/content/66/1/74

Chapter 2

Geological Background

Yasuto Itoh

Abstract

A tectonic graben within the central Kyushu Island, named the Hohi Volcanic Zone (HVZ), is buried by Plio-Pleistocene volcaniclastic materials. It is accompanied by affluent indicators of tensile stress and is often related to the N-S breakup of the continental crust of Kyushu. Such a hypothesis is, however, discordant with recent horizontal movements detected by global navigation satellite system (GNSS)-based analysis. The latest remote sensing data rather accord with the deformation field being provoked by westward indentation of the forearc sliver of southwest Japan according to intermittent dextral slips along the Median Tectonic Line (MTL). Although the MTL has a long-standing and complicated history of movements since the Cretaceous, it has been activated as a reverse, and then, right-lateral fault system under control of the convergence modes of the Philippine Sea Plate. The active MTL trace coincides with southern margin of the HVZ, and the right-stepping configuration of the dextral fault system resulted in the growth of a pull-apart basin around Beppu Bay.

Keywords: Kyushu, Hohi Volcanic Zone, Median Tectonic Line, Beppu Bay, Philippine Sea Plate, extrusion tectonics, pull-apart basin

1. Introduction

Located at an arc-arc junction, the island of Kyushu features a quite complicated evolutionary history accompanied with hyperactive volcanism and neotectonic movements. Among various controversies concerning the geology of this island, we selected two hot topics related to the book subject. They are, namely, the stress-strain state and fault architecture around central Kyushu.

2. Formation mechanism of the Hohi Volcanic Zone

The Hohi Volcanic Zone (HVZ) is a box-shaped tectonic graben buried by more than 5000 km^3 of volcaniclastic materials [1]. This volcanic province is studded with indications of prevailing tensile stress, such as dike swarms (see Figure 3 in Prologue), and often has been regarded as an eastern constituent of the active rift zone crossing central Kyushu. For example, Tada [2] advocated an N-S breakup of the continental crust based on 90-year-long repeated triangulation and trilateration surveys. However, the estimated extension rates significantly increase across the island from 1 mm/year in the HVZ (Pleistocene fault analysis [1]) to 14 mm/year around Shimabara Peninsula at the westernmost point of Kyushu (geodetic observation [2]). Such a lopsided motion seems to be contradictory to the working hypothesis of the rift valley as suggested by Itoh et al. [3].

An alternative mechanism of regional deformation in Kyushu was pointed out by Itoh and Takemura [4]. They took notice of lateral displacement of the forearc sliver of southwest Japan along the Median Tectonic Line (MTL). As shown in **Figure 1a**, the distribution of active strike-slip faults on the island basically follows a slip-line pattern provoked by long-standing westward crustal indentation. Recent horizontal fluctuation vectors for crustal movements detected by GNSS-based control points (**Figure 1b**) have agreed well with the hypothetical deformation field. Therefore, the strain in Kyushu is growing under the theory of extrusion just the same as structural buildup at a continental collision front.

As presented in **Figure 2a**, Itoh et al. [3] suggested that the HVZ has developed through two tectonic phases of the initial N-S extension (a-1) and succeeding pull-apart basin formation (a-2). Itoh and Inaba [5] suggested that a regime of increasing simple shear during the younger phase forced a northerly shift of the active MTL trace, which inevitably induced active migration of the depocenter. Reflecting this complex structural development, confined strong tension and compression are observed around the northwestern corner (rift-type monogenetic volcanism of Mt. Oninomi) and southeastern corner (subsurface inversion) of the basin, respectively (see **Figure 2a-2**). Spatiotemporal growth of the pull-apart sag is discussed in Chapter 4 based on a detailed seismic interpretation.

3. Fault architecture around Beppu Bay

The MTL has an exceptionally prolonged history of activity dating back to the Cretaceous, when a convergent margin along the northwestern Pacific suffered intense shear to form a regional detachment fault due to oblique subduction of oceanic plates [6]. It divides the southwestern Japan arc into the Inner and Outer Zones, which consist of the Paleozoic/Mesozoic accretionary complex intruded by voluminous igneous rocks and accreted/amalgamated bodies with high-pressure metamorphic rocks, respectively. In Kyushu, the Usuki-Yatsushiro Tectonic Line (**Figure 2b**) is a candidate for such a bisecting fault [7]. Its architectural context, however, remains controversial because later strong deformation brought about lateral transport of thrusted metamorphic terranes over the tectonic line [8–10].

After a dormant period in the Paleogene, the MTL was reactivated as a reverse, and then, right-lateral fault system under control of the fluctuant convergence of the Philippine Sea Plate [11, 12]. The dextral active fault along the southern coast of Beppu Bay is named the Saganoseki Fault (**Figure 2b** [13]). It is nearly continuous with the eastern part of the Oita-Kumamoto Tectonic Line, which coincides with the southern margin of the HVZ (see **Figure 2a-1**). On the opposite side of the volcano-tectonic graben, a swarm of superficial extensional features (Northern Marginal Fault Zone; **Figure 2b**) were confirmed by Shimazaki et al. [14]. Itoh et al. [3] interpreted the right-stepping configuration of the dextral MTL and the Beppu-kita Fault (**Figure 2b**) as confirming the emergence of a pull-apart basin bordered by large rollover structures on the Asamigawa and Central Beppu Bay Faults (**Figure 2b**).

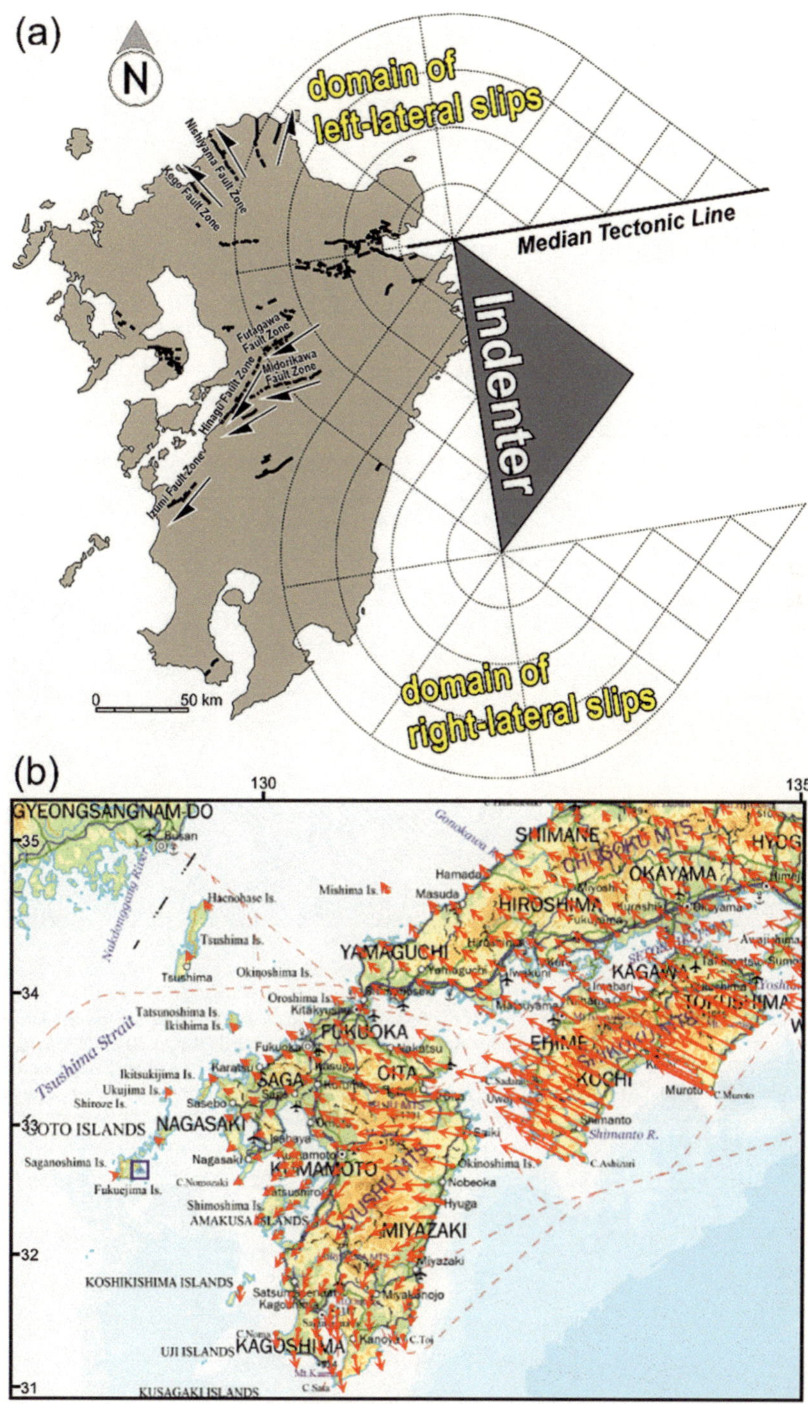

Figure 1.

(a) Crustal extrusion model of Kyushu after Itoh and Takemura [4]. (b) Recent horizontal fluctuation vectors (red arrows) of crustal movement detected by a GNSS-based control point (December 2017 to December 2022), as obtained from the website of the Geospatial Information Authority of Japan (https://mekira.gsi.go.jp/index.en.html). The fixed station on Fukue Island is shown by a blue square.

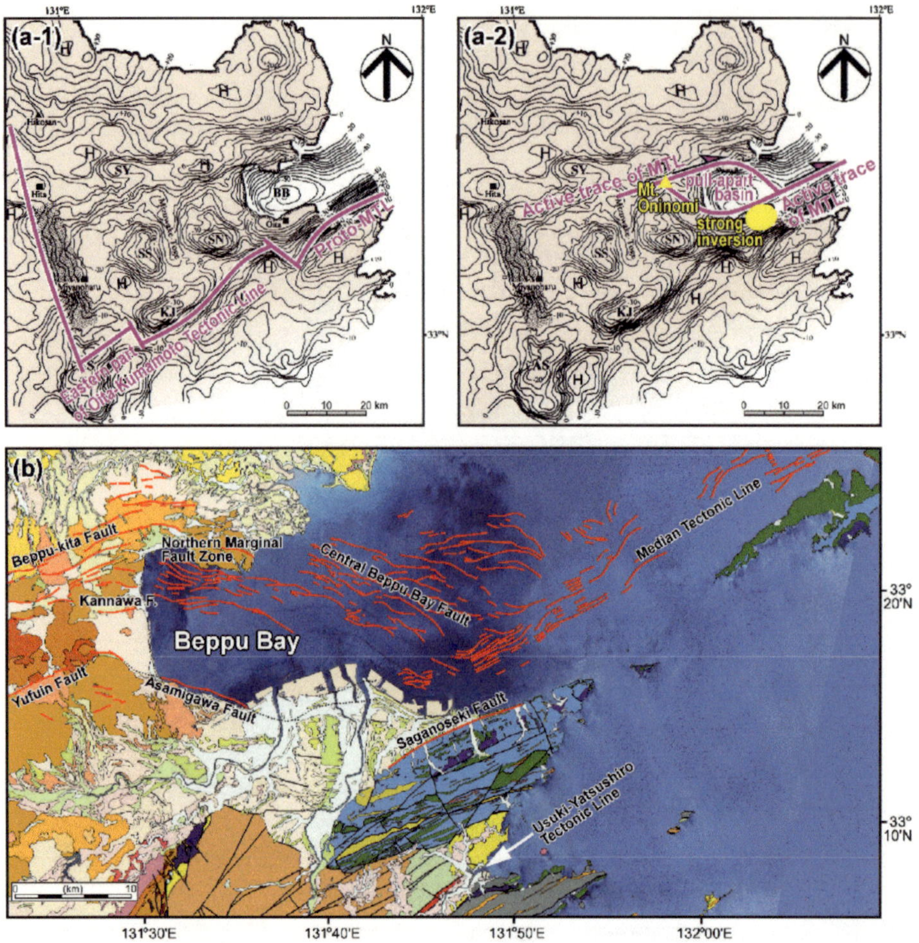

Figure 2.
(a) Development process of the HVZ since the Pliocene (1 = older, 2 = younger). The Bouguer gravity anomaly contours are after Itoh et al. [3]. Abbreviations for significant depressions are BB=Beppu Bay, SN=Shonai, KJ = Kuju, SY=Shin-Yabakei, SS=Shishimuta, and AS = Aso. The shaded zone shows a gravity slope on the western margin of the HVZ. (b) Neotectonic structural features around the study area are based on GeomapNavi, the geological map display system of the Geological Survey of Japan, AIST (https://gbank.gsj.jp/geonavi/geonavi.php#11.33.28137,131.78390).

DOI: http://dx.doi.org/10.5772/intechopen.112747

Author details

Yasuto Itoh
Osaka Metropolitan University, Osaka, Japan

*Address all correspondence to: yasutokov@yahoo.co.jp

References

[1] Kamata H. Volcanic and structural history of the Hohi volcanic zone, Central Kyushu, Japan. Bulletin of Volcanology. 1989;**51**:315-332

[2] Tada T. Spreading of the Okinawa trough and its relation to the crustal deformation in the Kyushu (2). Bulletin of Seismological Society of Japan (Zisin). 1985;**38**:1-12

[3] Itoh Y, Takemura K, Kamata H. History of basin formation and tectonic evolution at the termination of a large transcurrent fault system: Deformation mode of Central Kyushu, Japan. Tectonophysics. 1998;**284**:135-150

[4] Itoh Y, Takemura K. Mode of Quaternary crustal deformation of Kyushu Island, Japan. In: Shichi R, Heki K, Kasahara M, Kawasaki I, Murakami M, Nakahori Y, et al., editors. Proceedings of 8th International Symposium on Recent Crustal Movements. Kyoto: The Local Organizing Committee for the CRCM'93; 1994. pp. 407-411

[5] Itoh Y, Inaba M. Fault-Controlled Processes of Basin Evolution: A Case on a Longstanding Tectonic Line. New York: Nova Science Publishers, Inc.; 2019. p. 81

[6] Itoh Y, Takano O, Takashima R. Tectonic synthesis - a plate reconstruction model of the NW Pacific region since 100 Ma. In: Itoh Y, editor. Dynamics of Arc Migration and Amalgamation - Architectural Examples from the NW Pacific Margin. Rijeka: InTechOpen; 2017. DOI: 10.5772/67358

[7] Sano H. Geology of Japan, Kyushu-Okinawa. Tokyo: Asakura; 2010. p. 619

[8] Saito M, Miyazaki K, Toshimitsu S, Hoshizumi H. Geology of the Tomochi District with Geological Sheet Map at 1:50,000. Geological Survey of Japan, AIST: Tsukuba; 2005. p. 218

[9] Hoshizumi H, Saito M, Mizuno K, Miyazaki K, Toshimitsu S, Matsumoto A. Geological Map of Japan at 1:200,000, Oita (2nd Edition). Tsukuba: Geological Survey of Japan, AIST; 2015

[10] Saito M, Takarada S, Toshimitsu S, Mizuno K, Miyazaki K, Hoshizumi H, et al. Geological Map of Japan at 1:200,000, Yatsushiro and a Part of Nomo Zaki. Tsukuba: Geological Survey of Japan, AIST; 2010

[11] Itoh Y, Nagasaki Y. Crustal shortening of Southwest Japan in the late Miocene. The Island Arc. 1996;**5**:337-353

[12] Itoh Y, Iwata T, Takemura K. Three-dimensional architecture of the Median Tectonic Line in Southwest Japan based on detailed reflection seismic and drilling surveys. In: Itoh Y, editor. Evolutionary Models of Convergent Margins - Origin of their Diversity. Rijeka: InTechOpen; 2017. DOI: 10.5772/67434

[13] Geological Survey of Japan, AIST. Active Fault Database of Japan, 4 October 2016 Version. Tsukuba: Geological Survey of Japan, National Institute of Advanced Industrial Science and Technology; 2016. Available from: https://gbank.gsj.jp/activefault/index

[14] Shimazaki K, Nakata T, Chida N, Miyatake T, Okamura M, Shiragami H, et al. A preliminary report on the drilling project of submarine active faults beneath Beppu Bay, Southwest Japan, for longterm earthquake prediction. Active Fault Research. 1986;**2**:83-88

Chapter 3

Numerical Modeling

Shigekazu Kusumoto and Yasuto Itoh

Abstract

Numerical simulation to reproduce patterns of topography or subsurface structure is a useful technique for validating hypotheses regarding their forming processes and/or tectonics as their background. Here, the author describes an outline of dislocation modeling and several methods for applying it to reproduce topography or subsurface structure. Finally, it is reported that the formation processes of the Hohi Volcanic Zone proposed from geological viewpoints are validated by dislocation modeling.

Keywords: numerical modeling, dislocation modeling, pull-apart basin, restoration, Beppu Bay, Hohi Volcanic Zone

1. Introduction

Numerical modeling or numerical simulation is a useful technique for validating hypotheses and quantitatively evaluating phenomena. Analog experiments are also helpful for validating hypotheses and have often been used to analyze surface deformation or trishear zones caused by fault motions and to model caldera formation [1–5]. However, it is sometimes difficult to conduct laboratory experiments as analog experiments, depending on the scale of the structures or phenomena, because the similarity rule for the model, including physical properties, has to be considered. Additionally, issues with the quantitative analysis of the experimental results sometimes occur.

In contrast, numerical simulations are used for the validation and quantitative evaluation of various geological and geophysical phenomena, as they enable modeling on a realistic scale. Numerical simulation methods, such as dislocation modeling, finite element modeling, boundary element modeling, finite difference modeling, and discrete element methods, have frequently been used to solve solid earth science problems [6–14].

Pull-apart basins, the main structures in our study area, are known to form at the fault terminations of right-lateral, right-stepping or left-lateral, left-stepping fault zones [15]. Because the lateral fault motion creates subsidence at the fault terminations, pull-apart basins are formed at the terminations of the fault zone or fault arrangement area, as mentioned above, where the subsidence is superimposed, and they are widely distributed worldwide. Rodgers [6] was probably the first study to discuss the formation of pull-apart basins through numerical simulations using Chinnery's dislocation solution [16]. Currently, Okada's dislocation solutions [17, 18] are often employed to understand crustal deformation.

2. Dislocation modeling

Okada's dislocation solutions are closed analytical solutions that calculate the surface and internal deformation and strain fields due to shear and tensile faulting with an arbitrary dip in an elastic isotropic half-space.

In geodesy, these solutions have been employed for estimating fault parameters, such as fault displacement on fault surfaces (dislocation planes), fault positions, and dip, by inverse analysis of surface displacements obtained by GNSS, leveling, and other geodetic observations [19, 20]. However, in tectonics, we focus on the basic deformation patterns formed by fault motions and discuss whether a combination of fault motions can form the topography or subsurface structures estimated by geophysical prospecting [21, 22].

In restoration (reproduction) modeling of topography or subsurface structures, the analysis is conducted by forward modeling, in which the fault parameters are estimated through trial and error. The geological background and geophysical validities are considered in the modeling. The validity of the model for restoring topography or subsurface structures is determined by comparing the pattern of the calculated topography or subsurface structures with the pattern of the actual structures.

There is a technique in which large Poisson's ratios are assumed if the phenomena at geological time scales are modeled in elastic media [8, 22]. This method covers the influence of the cumulative deformation of the fault motion by the elastic constant, such as assuming a soft medium imitating fluid.

To reflect the multiple motions of active faults over geological timescales in dislocation modeling, Itoh et al. [21] introduced historical fault activities into the modeling by superimposing analytical solutions in which fault parameters for each fault motion are specified on the concerned fault. They attempted to reproduce the shape pattern of the Takayama Basin in central Japan by applying this technique and showed that it is a tectonic basin caused by the accumulated right-lateral motions of two active faults. In addition, it was revealed that this technique could explain not only the topography but also changes in the declination of thermoremanent magnetization. This technique has also been applied to validate the formation processes of tectonic basins distributed in central Hokkaido [22–24], and the fault motions and their combinations have been discussed.

Because these simulations focus on the shape pattern of the topography or subsurface structures, the calculated deformation field is often normalized by the absolute value of the maximum deformation.

3. Restoration of basement structure of the Hohi Volcanic Zone

A numerical simulation to reproduce the pattern of the basement structure of the Hohi Volcanic Zone (HVZ) [25], including Beppu Bay, was performed using dislocation modeling [26]. The basement structure was estimated by gravity analyses, considering surface geological information, drilling core data, and seismic prospecting data, and was shown to have three basins distributed along the Oita-Kumamoto Tectonic Line and Median Tectonic Line (MTL) [27]. Okada's dislocation plane [17] was employed in the modeling.

Following and simplifying the tectonic history of the HVZ, which Itoh et al. [28] proposed from a geological point of view, Kusumoto et al. [26] assumed that the basement structure was formed in three stages:

1. Formation of a half-graben (Stage I, Pliocene).

2. Formation of initial pull-apart basins due to the activation of right-lateral faults (Stage II, early Quaternary).

3. Growth of the pull-apart basins due to change in the active area of the right-lateral fault (Stage III, middle to late Quaternary).

In stage I, it was assumed that the Oita-Kumamoto Tectonic Line and the Kurume-Hiji Line [29] moved as normal faults under the strong N-S extension stress field. They formed a half-graben, which is the basic structure of the HVZ. In stage II, it was assumed that the MTL and the Kurume-Hiji Line moved as right-lateral faults under the E-W compression stress field. As a result, they formed an initial pull-apart basin (Beppu Bay and Shonai Basin). In stage III, the reduction in the active area of the MTL promoted the evolution of Beppu Bay. A series of numerical simulations showed that the tectonic models of the HVZ proposed by Itoh et al. [28] are also correct from a geophysical viewpoint.

In addition, Kusumoto et al. [30] assumed a two-layer model consisting of a basement and sedimentary layer in the HVZ. They attempted to calculate the gravity anomaly pattern caused by the basement structure reproduced by Kusumoto et al. [26]. They showed that gravity anomalies due to tectonic structures are mostly explained, except for the gravity anomaly caused by volcanic structures such as the Shishimuta caldera [31].

Author details

Shigekazu Kusumoto[1*] and Yasuto Itoh[2]

1 Institute for Geothermal Sciences, Kyoto University, Oita, Japan

2 Osaka Metropolitan University, Osaka, Japan

*Address all correspondence to: kusu@bep.vgs.kyoto-u.ac.jp

References

[1] Tsuneishi Y. Geological and experimental studies on mechanism of block faulting. Bulletin of Earthquake Research Institute University of Tokyo. 1978;**53**:173-242

[2] Komuro H. Experiments on caldron formation: A polygonal cauldron and ring fractures. Journal of Volcanology and Geothermal Research. 1987;**31**:139-149

[3] Acocella V, Cifelli F, Funiciello R. Analogue models of collapse calderas and resurgent domes. Journal of Volcanology and Geothermal Research. 2000;**104**:81-96

[4] Roche O, Druitt TH, Merle O. Experimental study of caldera formation. Journal of Geophysical Research. 2000;**105**:395-416

[5] Jin G, Groshong RH Jr. Trishear kinematic modeling of extensional fault propagation folding. Journal of Structural Geology. 2006;**28**:170-183

[6] Rodgers DA. Analysis of pull-apart basin development produced by en echelon strike-slip faults. Special Publications, International Association of Sedimentologists. 1980;**4**:27-41

[7] Bertoluzza L, Perotti CR. A finite-element model of the stress field in strike-slip basins: Implications for the Permian tectonics of the southern Alps (Italy). Tectonophysics. 1997;**280**:185-197

[8] Katzman R, ten Brink US, Lin J. Three-dimensional modeling of pull-apart basins: Implications for the tectonics of the Dead Sea Basin. Journal of Geophysical Research. 1995;**100**:6295-6312

[9] Du Y, Aydin A. Shear fracture patterns and connectivity at geometric complexities along strike-slip faults. Journal of Geophysical Research. 1995;**100**:18093-18102

[10] Finch E, Hardy S, Gawthorpe RL. Discrete element modelling of extensional fault-propagation folding above rigid basement fault blocks. Basin Research. 2004;**16**:489-506

[11] Hardy S, Finch E. Discrete-element modelling of detachment folding. Basin Research. 2005;**17**:507-520. DOI: 10.1111/j.1365-2117.2005.00280.x

[12] Panien M, Buiter SJH, Schreurs G, Pfiffner OA. Inversion of a symmetric basin: Insights from a comparison between analogue and numerical experiments. In: Buiter SJH, Schreurs G, editors. Analogue and Numerical Modelling of Crustal-Scale Processes. Vol. 253. London: Geological Society, London, Special Publications; 2006. pp. 253-270

[13] Yamada Y, Baba K, Miyakawa A, Matsuoka T. Granular experiments of thrust wedges: Insights relevant to methane hydrate exploration at the Nankai accretionary prism. Marine and Petroleum Geology. 2014;**51**:34-48. DOI: 10.1016/j.marpetgeo.2013.11.008

[14] Kusumoto S, Itoh Y, Takemura K, Iwata T. Displacement fields of sedimentary layers controlled by fault parameters: The discrete element method of controlling basement motions by dislocation solutions. Earth Science. 2015;**4**:89-94

[15] Aydin A, Nur A. Evolution of pull-apart basins and their scale independence. Tectonics. 1982;**1**:91-105

[16] Chinnery MA. The deformation of the ground around surface faults.

Bulletin of Seismological Society of America. 1961;**51**:355-372

[17] Okada Y. Surface deformation due to shear and tensile faults in a half-space. Bulletin of Seismological Society of America. 1985;**75**:1135-1154

[18] Okada Y. Internal deformation due to shear and tensile faults in a half-space. Bulletin of seismological society of. America. 1992;**82**:1018-1040

[19] Lasserre C, Peltzer G, Crampé F, Klinger Y, Van der Woerd J, Tapponnier P. Coseismic deformation of the 2001 Mw=7.8 Kokoxili earthquake in Tibet, measured by synthetic aperture radar interferometry. Journal of Geophysical Research. 2005;**110**. Article ID: B12408. DOI: 10.1029/2004JB003500

[20] Hotta K, Kusumoto S, Takahashi H, Hayakawa YS. Deformation source revealed from leveling survey in Jigokudani valley, Tateyama volcano, Japan. Earth, Planets and Space. 2022;**74**:32. DOI: 10.1186/s40623-022-01593-7

[21] Itoh Y, Kusumoto S, Furubayashi T. Quantitative evaluation of the Quaternary crustal deformation around the Takayama basin, Central Japan: A paleomagnetic and numerical modeling approach. Earth and Planetary Science Letters. 2008;**267**:517-532. DOI: 10.1016/j.epsl.2007.11.062

[22] Tamaki M, Kusumoto S, Itoh Y. Formation and deformation process of the late Paleogene sedimentary basins in the southern Central Hokkaido, Japan: Paleomagnetic and numerical modeling approach. Islands Arc. 2010;**19**:243-258

[23] Itoh Y, Kusumoto S, Maeda J. Cenozoic basin-forming processes along the northeastern margin of Eurasia: Constraints determined from geophysical studies offshore of Hokkaido, Japan. Journal of Asian Earth Sciences. 2009;**35**:27-33

[24] Kusumoto S, Itoh Y, Takano O, Tamaki M. Numerical modeling of sedimentary basin formation at the termination of lateral faults in a tectonic region where fault propagation has occurred. In: Itoh Y, editor. Mechanism of Sedimentary Basin Formation - Multidisciplinary Approach on Active Plate Margins. Rijeka: InTechOpen; 2013. pp. 273-304. DOI: 10.5772/56558

[25] Kamata H. Volcanic and structural history of the Hohi volcanic zone, Central Kyushu, Japan. Bulletin of Volcanology. 1989;**51**:315-332

[26] Kusumoto S, Takemura K, Fukuda Y, Takemoto S. Restoration of the depression structure at the eastern part of Central Kyushu, Japan, by means of the dislocation modeling. Tectonophysics. 1999;**302**:287-296

[27] Kusumoto S, Fukuda Y, Takemoto S, Yusa Y. Three-dimensional subsurface structure in the eastern part of the Beppu-Shimabara Graben Kyushu, Japan, as revealed by gravimetric data. Journal of the Geodetic Society of Japan. 1996;**42**:167-181

[28] Itoh Y, Takemura K, Kamata H. History of basin formation and tectonic evolution at the termination of a large transcurrent fault system: Deformation mode of Central Kyushu, Japan. Tectonophysics. 1998;**284**:135-150

[29] Kido M. Tectonic history of the north-western area of 'Kuju-Beppu Graben'. Memoirs of the Geological Society of Japan. 1993;**41**:107-127 (in Japanese with English abstract)

[30] Kusumoto S, Fukuda Y, Takemura K. A distinction technique between volcanic

and tectonic depression structures based on the restoration modeling of gravity anomaly: A case study of the Hohi volcanic zone, Central Kyushu, Japan. Journal of Volcanology and Geothermal Research. 1999;**90**:183-189

[31] Kamata H. Shishimuta caldera, the buried source of the Yabakei pyroclastic flow in the Hohi volcanic zone, Japan. Bulletin of Volcanology. 1989;**51**:41-50

Chapter 4

Ultra-High-Resolution Seismic Surveys: 3D Sea Trial at Beppu Bay

Yousuke Teranishi, Takeshi Kozawa, Hitoshi Tsukahara, Fumitoshi Murakami, Yasuto Itoh and Shigekazu Kusumoto

Abstract

A 3D high-resolution seismic (3D-HRS) survey was conducted to clearly reveal and map faults and fractures in a shallow-water region of Beppu Bay, Japan. The 3D-HRS was conducted using a dense array of six short streamer cables combined with a single GI gun, which generate the primary pulse and create the main bubble with generator and collapse the main bubble with injector. This high-frequency seismic source and high-density seismic source/receiver point distribution achieved a significant improvement in seismic imaging resolution. Because the 3D-HRS system is compact and lightweight and can be operated by small vessels, its effectiveness and necessity have been demonstrated, especially in shallow coastal waters, where conventional seismic survey using large ships and long streamer cables is difficult. For seismic data processing, pre-stack noise attenuation, de-ghosting, multiple removal, and acquisition footprint removal had essential roles in enhancing seismic imaging quality. Compared to existing 2D seismic survey technology, the 3D-HRS achieved much higher resolution and delineated highly detailed features of the seafloor and subsurface. Following the seismic processing sequence, similarity and thinned fault likelihood attribute workflows were applied to detect and visualize faults and fractures within the 3D-HRS volume. These seismic attributes revealed a network of broadly distributed faults and fractures along an active fault system in Beppu Bay.

Keywords: high-resolution seismic, shallow water, seismic data acquisition, seismic data processing, seismic attribute

1. Introduction

In recent years, the importance of geotechnical surveys to identify and evaluate geological risks related to seafloor topography and seafloor characteristics in shallow coastal waters has been increasing. Recently, the development of offshore wind power generation projects has been promoted by the public and private sectors, and the establishment of techniques to properly identify, evaluate, and manage geological seabed risks for the structural foundations of wind turbines has become an urgent issue (e.g., [1]). Because Japan exists in a variable zone where plate boundaries exist, it has unique geohazards different from those of other areas, such as Europe. Therefore, it is necessary to take appropriate responses to deal with the geological risks, especially in coastal waters, of active faults, earthquakes, submarine landslides, volcanoes, shallow gases, and sediment fills in old river channels.

Seismic surveys have been applied not only to marine resource surveys but also to active fault surveys, geohazard surveys, and monitoring surveys. In recent years, the development of high-resolution seismic (HRS) survey technology, which consists of high-frequency seismic sources and short streamer cables with a high-density channel arrangement, has progressed with the aim of understanding the detailed subsurface structure in shallow areas below the seafloor. These HRS techniques use a seismic source frequency of 200 to 400 Hz, and ultra-high-resolution seismic (UHRS) surveys use a seismic source frequency of 2 to 3 kHz. The HRS/UHRS technology has been put into practical use and achieved a significant improvement in vertical and horizontal resolution by using high-frequency seismic sources and high-density seismic source and receiver point distributions. Because the HRS/UHRS system is compact and lightweight and can be operated by small vessels, its effectiveness and necessity have been demonstrated, especially in coastal shallow waters where conventional seismic surveys using large ships and long streamer cables are difficult (e.g., [2, 3]). The vertical resolution and detection depth range in offshore seismic reflections largely depend on the frequency band of the seismic source used, with higher-frequency sources having higher vertical resolution but lower penetration due to greater energy attenuation during seismic wave propagation. Due to this trade-off between vertical resolution and detection depth range in HRS/UHRS systems, they are often used in different ways depending on the survey target. Suda et al. [4] reported improvements of vertical resolution and detection depth range by multi-scale exploration using multiple seismic sources with different frequency bands simultaneously.

JGI, Inc., has developed lightweight short streamer cables to make improvements by reducing cable towing noise, improving real-time positioning accuracy, and controlling cable position. After the 3D UHRS survey in the Yatsushiro Sea in Kumamoto Prefecture [3], which was conducted as part of the comprehensive active fault survey by the Japan Ministry of Education, Culture, Sports, Science and Technology in 2016, the HRS/UHRS system with independent recording type streamer cables (i.e., an autonomous cable system) is now commercially available. In this chapter, we show the results of a demonstration test conducted in Beppu Bay in an effort to understand the depth and shape of an engineered foundation, which is important for offshore wind power generation projects, and to visualize and extract geohazard factors such as faults, shallow gas, and soft layers. Then, we discuss the utility of high-resolution 3D seismic surveys in shallow coastal waters using the HRS/UHRS system.

2. 3D high-resolution seismic survey data acquisition system

The 3D-HRS data acquisition system is a compact system developed to obtain a detailed understanding of the geological structure of the shallow subsurface below the seabed. The system consists of short streamer cables, small high-frequency seismic sources, and onboard equipment. The system configuration and main technical specifications of the 3D-HRS data acquisition system are shown in **Figure 1** and **Table 1**, respectively. Closely spaced shot points and receivers allow high-density

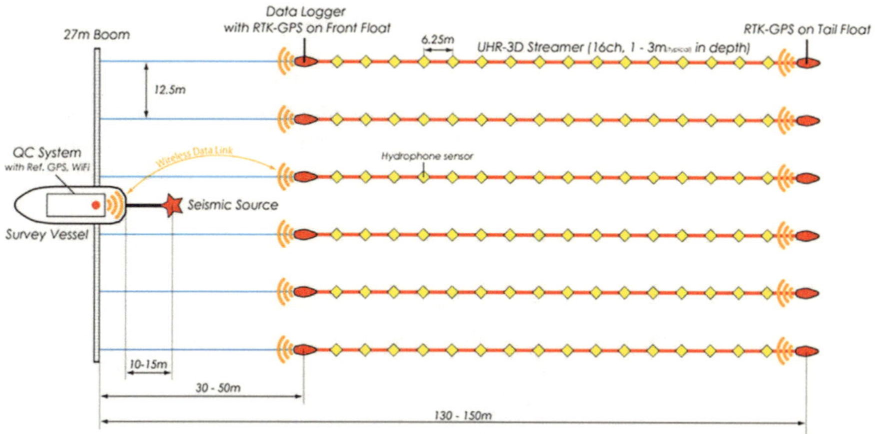

Figure 1.
Schematic configuration of 3D-HRS data acquisition system.

Streamer cable	**ACS (JGI, Inc.)**
Number of channels/line	16 ch
Channel spacing	6.25 m
Total length of streamer	100 m
Outer diameter	20 mm
Sensitivity of hydrophone	–201 dB re: 1 V/uPa
Frequency response	5 Hz–10 kHz (±3 dB)
Autonomous data recording system	**JGI-HR3D-2019 (JGI, Inc.)**
Number of input channels	16 ch
A/D converter	24 bit ΔΣ
Sampling frequency	1 kHz/4 kHz/10 kHz
Dynamic range	109 dB
Preamplifier gain (selectable)	xl/x4/x8/xl6
Navigation system	**vNAV (JGI, Inc.)**
Navigation	D-GPS, RTK-GPS
Shooting interval	Constant distance intervals/constant time intervals

Table 1.
Specifications of the 3D-HRS system.

seismic records to be acquired around the seismic spread, providing a high-resolution, accurate understanding of the 3D subsurface structure.

A high-resolution 3D seismic survey was conducted using the 3D-HRS data acquisition system with a small vessel having a gross tonnage of approximately 300 tons. Several short streamer cables around 100 m long were deployed using crane booms installed on either side near the stern of the vessel. The system was designed to efficiently perform high-resolution 3D seismic surveys over a range of several kilometers to several tens of kilometers at low cost. The main features of this system are described below.

2.1 Autonomous cable system

The autonomous cable system is an analog streamer cable with 16 highly sensitive hydrophone sensors placed at 6.25-m intervals. The cable is thin (20 mm in diameter) and lightweight, allowing it to be deployed and retrieved manually (no streamer winch required) and requiring only a small deck space on the survey vessel.

2.2 Cable towing system using crane booms

In areas with heavy vessel traffic, such as bays and coastal areas, it is necessary to conduct surveys in a maneuverable manner. Therefore, instead of using large paravanes, as commonly used in marine 3D seismic surveys, the streamer cables are towed from crane booms (less than 30 m long) extending from both sides of the stern of the vessel. The system has a very small turning radius (300–400 m) and can successfully tow multiple short streamer cables in a row, even when currents and winds prevent the vessel from maintaining a prescribed speed.

2.3 Cable depth and position control

The streamer cables are designed to have neutral buoyancy in seawater, and the towing depth is set by depth control devices located under the front and tail floats. The towing depth is observed with a depth gauge attached near the middle of each cable, and the weight of the cable is adjusted to maintain the specified towing depth as much as possible. This system does not use any active towing depth control device, such as a bird, to reduce cable towing noise. The steering control devices are mounted on the tail floats that try to maintain the streamer cables in certain positions. GPS devices are mounted on all floats to measure the exact positions of the receivers and the seismic source. It is also the role of the GPS devices to provide accurate time synchronization of all observation systems.

2.4 Data communication between the cables and survey vessel

A data logger is mounted on the front float of each cable, and signals from the hydrophone sensors and cable depth gauge are converted from analog to digital and stored in the data logger. Seismic data and GPS positioning information stored in the data logger and control signals to the steering control device on the tail float are transmitted *via* Wi-Fi in real time to and from the observation room on the survey vessel.

2.5 Quality control of seismic records

The onboard equipment consists of the 3D-HRS recording system developed by JGI, Inc., which collects and displays seismic data transmitted from the data loggers *via* Wi-Fi. A quality control (QC) system manages the quality of seismic records and the navigation system (vNAV system developed by JGI, Inc.), which navigates the survey vessel and controls the shot-timing of the seismic source. The real-time QC of seismic records and the position of the seismic spread is performed in the observation room to detect anomalies in each system and makes corrections to meet survey requirements by clients and/or survey designers.

3. Sea trial evaluation of the 3D-HRS survey

3.1 Survey area

We conducted a sea trial of the 3D-HRS survey system in Beppu Bay, Oita Prefecture, which is located at the western end of the Median Tectonic Line (MTL) and has many active fault systems that are a mixture of strike-slip faults and normal faults (e.g., [5]). To evaluate the activity of this fault zone, it is important to precisely understand the fault locations, distribution geometry, and displacement in the upper part of the Holocene deposits. In this 3D-HRS trial, data acquisition specifications were designed to obtain seismic records that contribute to the evaluation of the activity of subsurface active faults, with a vertical resolution of approximately 2 m and a maximum investigation depth of 300 m or more below the seafloor. The water depth in the survey area was 30 to 50 m.

A marine 2D seismic survey (9 survey lines with a total length of 138 km) was conducted by the Faculty of Science, Kyoto University, in 1989 to investigate the deep subsurface structure of Beppu Bay, and Yusa et al. [6] obtained groundbreaking insights into the geometry and tectonic motion of the MTL and the formation process of Beppu Bay, as well as the deep subsurface structure. Subsequently, in a 3-year national project starting in fiscal year 2014, new survey data were acquired for highly accurate strong-motion prediction of this fault zone and a comprehensive study, including the existing data, was carried out to update the fault distribution and deep structure from the coastal sea to land areas around Beppu Bay [7, 8]. However, deformation of the sedimentary layer due to fault movement diffuses and bifurcates near the ground surface and the seafloor. Therefore, the development of 3D high-resolution seismic survey technology has been identified as necessary for advanced evaluation of long-term earthquake prediction.

3.2 Data acquisition

The 3D-HRS sea trial was conducted in March 2020 in the southeastern part of Beppu Bay, the gateway to Beppu Port, Oita Port, and the steel and oil refinery complexes located in the Oita coastal industrial zone. The survey area was a rectangle of 6 km by 3 km (18 km^2), as shown in **Figure 2**. The area is an important marine traffic route for passenger vessels, iron ore and other raw material vessels, oil tankers, LNG carriers, and other vessels and is also a good fishing ground. Many ships, including pleasure boats and fishing boats, pass through this area day and night throughout the year in the survey area. Therefore, a highly maneuverable offshore 3D seismic survey system was required to facilitate the survey. The inline direction in this seismic survey was set to be orthogonal to the MTL (north-northwest to south-southeast) to facilitate comparison with the existing 2D seismic surveys (inline direction: long axis direction of the rectangular survey area). **Table 2** shows the data acquisition specifications.

In this survey, six short streamer cables approximately 100 m long were towed at 12.5-m intervals. Because there was only one seismic source in the survey, the common midpoint distribution area (swath width) in the crossline direction (orthogonal to the inline direction) obtained with a single survey line was 37.5 m, and the bin size was 3.125 m (inline) by 6.25 m (crossline). The width of the survey area in the crossline direction was 3 km, so the number of planned seismic survey lines was 81.

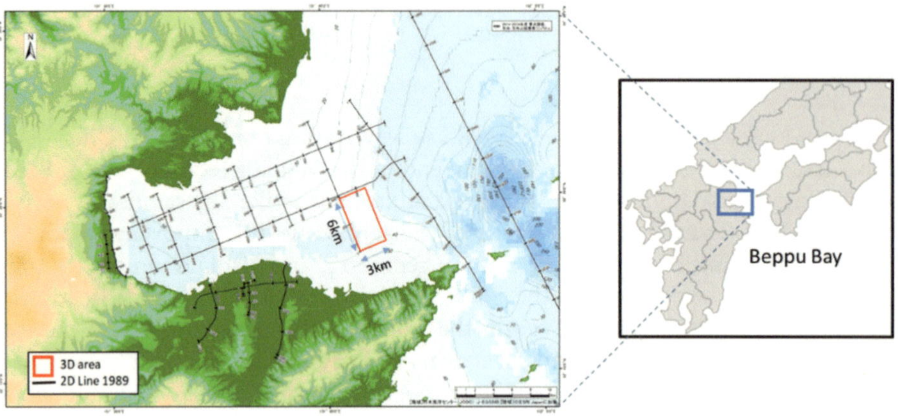

Figure 2.
Map of the survey area. The red square indicates the area of the 3D-HRS survey. The total area is approximately 18 km^2. Black lines indicate the lines of the seismic survey conducted by Kyoto University in 1989. All lines were reprocessed with state-of-the-art techniques [7].

Streamer cable	**ACS (JGI, Inc.)**
No. of streamers	6
Line spacing	12.5 m
Streamer depth	1.5 m
Total number of channels	96 ch
Channel spacing	6.25 m
Seismic source	**GI-Gun (Sercel)**
Capacity	150 cu.in. (45 + 105)
Shooting interval	6.25 m
Source depth	3.0 m
Recording system	**JGI-HR3D-2019 (JGI, Inc.)**
Sampling rate	0.25 ms
Record length	3.0 s
Bin size (Inline × Xline)	3.125 m × 6.25 m
Nominal number of folds	8

Table 2.
Specifications of the data acquisition trial in Beppu Bay.

The coordinates of shot points of the seismic source were directly measured with high precision by the GPS device mounted on the source float. The coordinates of the receiving points were calculated by interpolating the highly accurate positions measured by the GPS devices mounted on the front and tail floats attached to both ends of the streamer cables.

During seismic data acquisition, feathering of the streamer cables (deviations of the towing streamers from the planned positions) was observed in real time and the streamer spread was automatically optimized by the steering devices mounted on

Figure 3.
Photo of towed streamers taken from the sky.

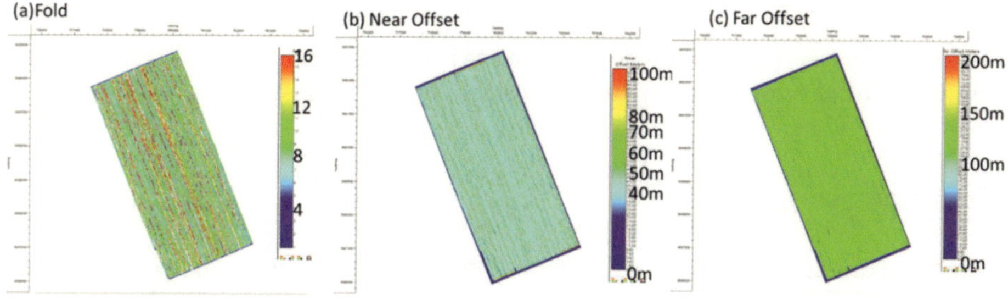

Figure 4.
Fold and offset distribution. (a) Number of folds for each bin (3.125 m inline, 6.25 m crossline). (b) Near-offset (40–50 m) distribution for each bin. (c) Far-offset (100–120 m) distribution for each bin. Both near- and far-offset data were uniformly acquired.

the tail floats of the streamer cables so as to maintain the specified towing interval for the streamer cables (**Figure 3**). The spatial distributions of the number of folds, the near offsets, and the far offsets for each bin obtained in the survey are shown in **Figure 4**.

The onboard QC of the survey data included real-time monitoring and confirmation of seismic records, amplitude spectra, near-trace records, fold maps, and noise levels (noise RMS amplitude values for every channel) at each seismic line, as well as checking the quality and consistency of the source waveforms acquired by the hydrophone sensor installed near the seismic source (**Figure 5**). The actual distribution of reflection points was found to be slightly spread out from the planned positions due to deflection of the towed streamers from the planned survey lines caused by sea conditions, obstructions at sea, and currents. Because missing reflection points in the acquisition coverage can contribute to significant compromise of the imaging quality of the seismic records, the onboard QC checked the number of stacking folds in each bin in real time during data acquisition, and if a certain number of zero-fold bins were found in the crossline direction due to deviations of the reflection point distribution, survey lines (infill seismic survey lines) were added to fill the gaps. The final number of survey lines was 112, of which 81 were planned and 31 were infill. It

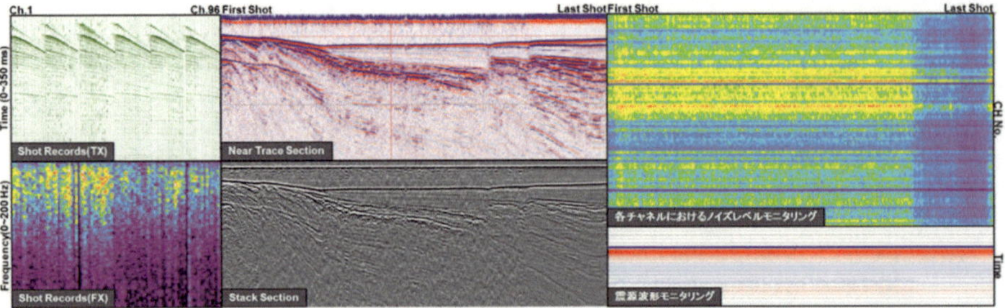

Figure 5.
Examples of onboard QC. Clear reflection events were observed in shot records, near trace sections, and in stack sections. Background noise and near-field source wavelets were monitored and used for system error detection in the survey.

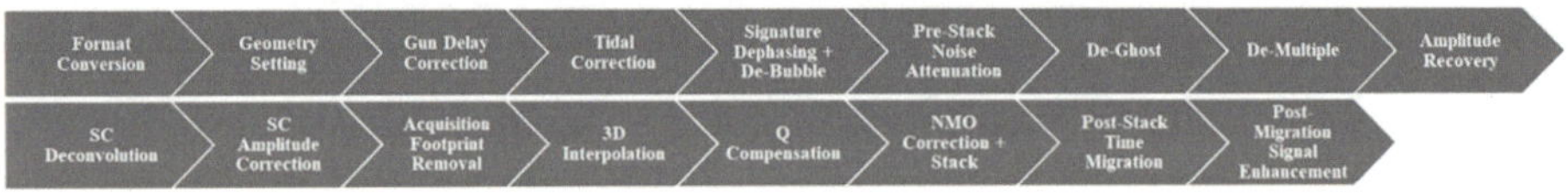

Format Conversion → Geometry Setting → Gun Delay Correction → Tidal Correction → Signature Dephasing + De-Bubble → Pre-Stack Noise Attenuation → De-Ghost → De-Multiple → Amplitude Recovery

SC Deconvolution → SC Amplitude Correction → Acquisition Footprint Removal → 3D Interpolation → Q Compensation → NMO Correction + Stack → Post-Stack Time Migration → Post-Migration Signal Enhancement

Figure 6.
Flow chart of 3D-HRS data processing.

took approximately 1.5 hours to observe one survey line of 6 km and a total of 9 days to complete the survey, including weather standby and infill surveys.

4. Data processing

The 3D-HRS data, which have enhanced resolution and improved imaging accuracy, require an adaptive seismic processing sequence with attention to detail in order to maintain the higher frequencies in the original seismic field records. Especially for 3D-HRS in shallow waters, short-period multiples and strong swell noise due to shallowly towed streamers are critical issues in processing. Positioning errors, varying tide, and wave conditions or velocity changes in the water column cause imaging artifacts due to severe static shifts. Here, we focus on these issues and introduce the seismic processing sequence summarized in **Figure 6**. The main seismic processing steps are described below.

4.1 Pre-stack noise attenuation

The acquired seismic data contained various types of noise, such as sudden strong amplitude noise, coherent noise that may have been caused by surrounding vessels, and noise that may have been generated by the survey vessel or front and tail floats. Therefore, we applied a noise suppression process that combines various methods in consideration of the noise characteristics. The applied noise suppression techniques are an FX (frequency vs. space) edit filter (burst noise suppression), FX prediction filter (random noise suppression), and velocity filter (linear noise suppression), which are applied in both the shot and channel domains. **Figure 7a** and **b** show seismic stack sections and amplitude spectra before and after the noise suppression process. These figures show that the noise components on the seismic stack section were effectively suppressed, improving the quality of the stack section.

DOI: http://dx.doi.org/10.5772/intechopen.112850

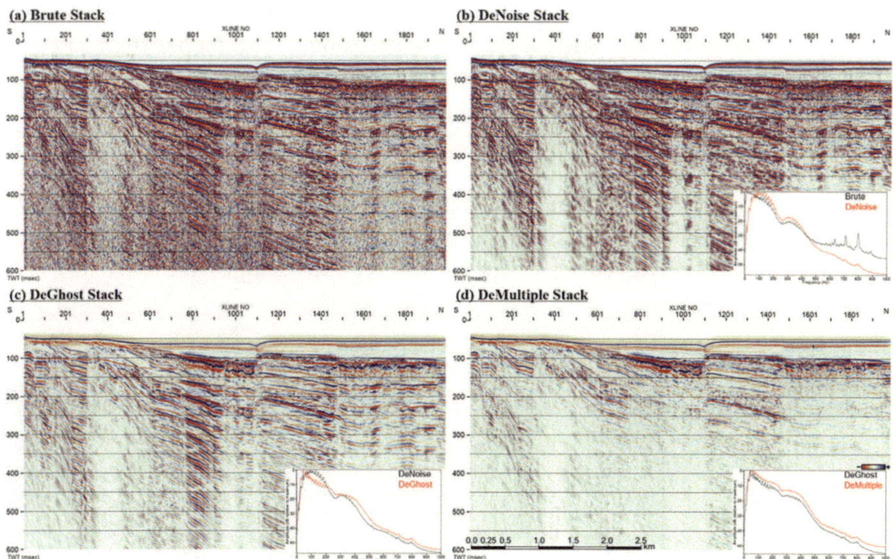

Figure 7.
Examples of seismic data processing along a 2D inline section and their frequency amplitude spectra. (a) Brute stack. (b) De-noised stack. (c) De-ghosted stack. (d) De-multipled stack. The pre-stack noise attenuation reduced several sources of noise and enhanced laterally continuous events. In addition, the zero-offset de-multiple method effectively suppressed the apparent reverberation below the seafloor reflection.

4.2 Multiple removal

Because water depths within the survey area ranged from 30 to 50 m, short-period multiple reflections dominated the seismic section and interfered with the primary reflections. The 3D-HRS system has a short maximum offset, which limits the effectiveness of surface-related multiple elimination and high-resolution parabolic Radon transform methods, which are typical multiple suppression methods used in conventional marine seismic surveys. To process the Beppu data, therefore, a zero-offset de-multiple (ZOD) technique based on approximate 1D modeling of multiple components was implemented. The ZOD was conducted in three steps. The first step applied time-shifts to seismic traces using the two-way travel time of the seafloor reflection for modeling and suppressing water-layer-related multiples. The second step was an auto-convolution of each trace for modeling and suppressing other long-period free-surface multiples. The third step was a long-gap deconvolution with the water depth as the gap distance to suppress residual multiples. **Figure 7c** and **d** show seismic stack sections and amplitude spectra before and after the ZOD process. These figures show that the quality of stack sections was improved by effectively suppressing multiples. The ripples in the amplitude spectrum caused by multiples were also reduced.

4.3 Velocity analysis

Generally, velocity analysis is difficult with 3D-HRS data due to the limitation of the maximum offset. However, because water depths within the survey area were shallow (from 30 to 50 m), there was sufficient variation in the angle of incidence to reflection surfaces up to about 200 ms below the seafloor. Consequently, the velocity analysis was possible up to about 150 m below the seafloor. Velocity analysis examples with a constant velocity stack, normal moveout common midpoint gather, and velocity semblance are shown in **Figure 8**. The velocity trend in the deeper than 150 m was estimated from other seismic data acquired by ocean-bottom nodes.

4.4 Footprint suppression

In the time slice shown in **Figure 9a**, significant acquisition footprints and missing traces along the sail line direction are observed. In addition to residual tidal changes after tidal correction, water column statics and irregular changes in cable depth caused

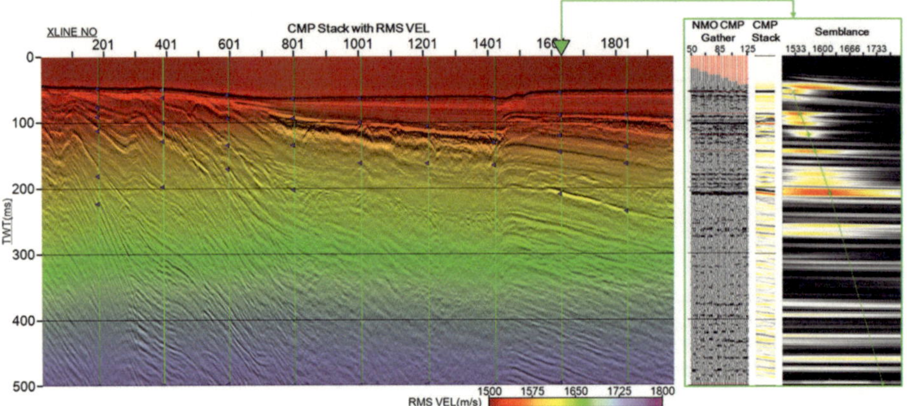

Figure 8.
Examples of the seismic velocity analysis. The velocity semblance spectrum indicates velocity trends until 300 ms, even though the offset range is limited to about 30–150 m.

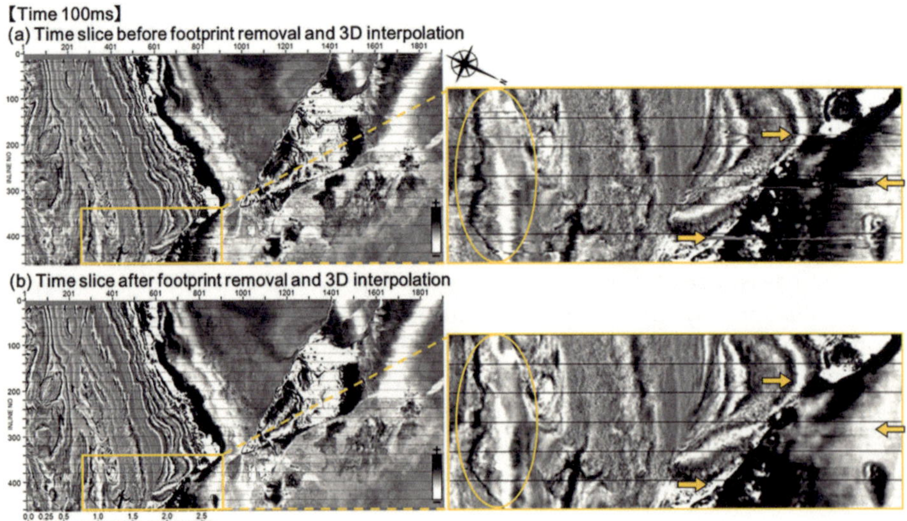

Figure 9.
(a) Enlargement of original time slice (time = 100 ms). The yellow circle and arrows indicate, respectively, acquisition footprints and missing traces. (b) The same slice after footprint suppression and 3D interpolation. These processes are shown to be effective for enhancing seismic images.

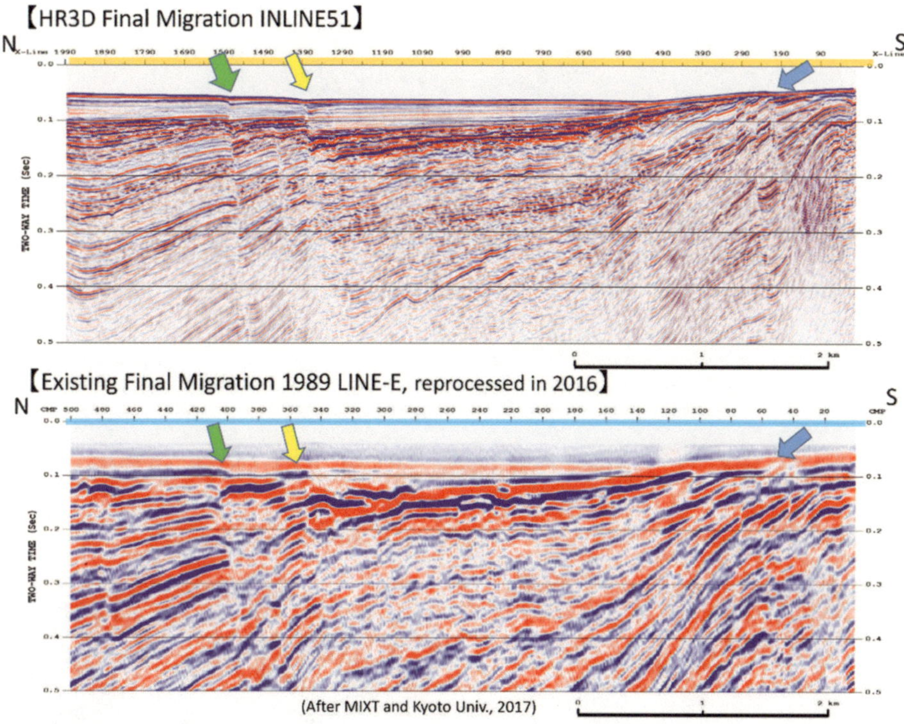

Figure 10.
(Top) Inline section of the final 3D-HRS dataset. (Bottom) Section of the existing 2D seismic record re-processed in 2016. Colored arrows indicate typical subsurface discontinuities highlighted in the 3D-HRS section.

severe static shifts. Such footprints become more pronounced as the vertical resolution of the seismic data improves, so footprint suppression is an important processing step for high-resolution surveys. Footprint suppression was applied to compensate for differences in reflection travel time due to the above factors and to suppress small spatial discontinuities in the reflection events. The footprint suppression included residual statics correction, trim statics, and smoothing of the seafloor reflections. After the footprint suppression, a structure dip-based 3D trace interpolation was conducted to correct irregular distributions of folds and offsets. **Figure 9a** and **b** show time slices before and after footprint suppression. The **Figure 9b** shows that the irregular changes in reflection events at **Figure 9a** were suppressed, and the continuity of reflection events was greatly improved. In addition, small gaps in acquisition coverage were successfully filled in (**Figure 9b**).

The vertical resolution in the shallow subsurface of the final migrated 3D-HRS volume was approximately 2.5 m based on the dominant frequency (150 Hz) and estimated P-wave velocity (1500 m/s). Compared to existing 2D seismic sections, 3D-HRS has much higher resolution. In particular, 3D-HRS can be seen to effectively extract spatial discontinuities, such as faults and fractures (**Figure 10**). The 3D view of the final migration results (**Figure 11**) shows the detailed subsurface shallow 3D structure with seafloor topography. In the vertical section and time slices, a spatially continuous fault distribution is clearly observed. This suggests that the 3D fault plane tracking using the 3D-HRS migration volume is more reliable than a 2D seismic survey.

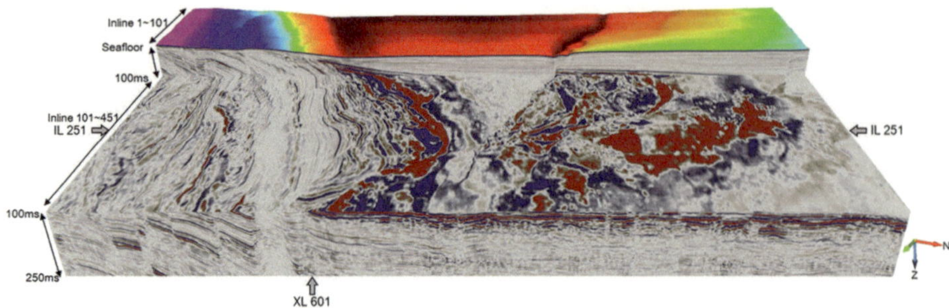

Figure 11.
3D-HRS post-stack time migration volume. In addition to the detailed 3D subsurface structure, the seafloor topography is also mapped with high accuracy. A spatially continuous fault distribution is clearly observed, indicating that 3D fault plane tracking using the 3D-HRS migration volume is more reliable than 2D seismic surveys.

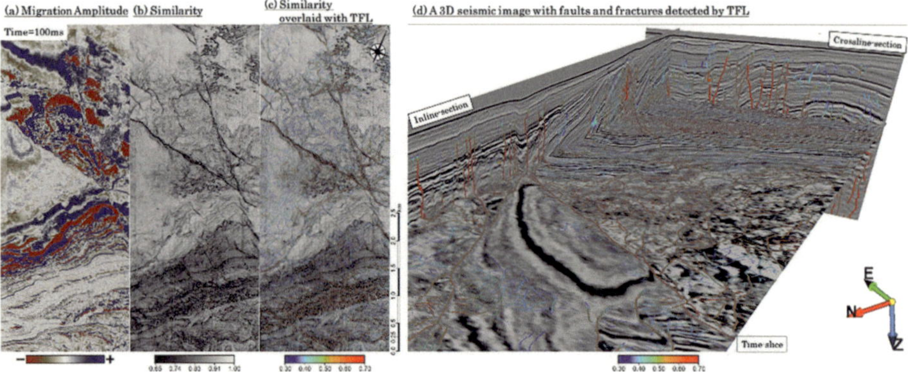

Figure 12.
Time slices at 100 ms of (a) input migrated volume, (b) similarity, and (c) similarity overlaid with TFL. The similarity attribute represents the discontinuous features of structural and stratigraphic trends. The TFL effectively highlights the fractures and faults within the similarity image. (d) 3D seismic image displayed with the faults and fractures detected by TFL. TFL successfully detects complicated subsurface 3D faults and fractures.

5. Seismic attribute analysis

Seismic attribute calculations are essential steps in the 3D seismic interpretation workflow. Seismic attributes can accelerate and provide quantitative justification for the interpretation of geologic features. We calculated the similarity and thinned fault likelihood (TFL) attributes of the post-stack time-migrated 3D-HRS volume for mapping of faults and fractures.

5.1 Similarity

Similarity is an attribute that indicates spatial coherency, with a value between 0 (discontinuity) and 1 (continuity). It is an indicator of structural continuity and directionality and is particularly useful for extracting spatial discontinuities such as faults and fractures.

5.2 Thinned fault likelihood

The TFL is likewise a useful attribute for fault and fracture extraction and is expected to indicate the probability of their presence, as well as provide highly accurate fault and fracture imaging [9]. **Figure 12a-c** show time-slice examples at 100 ms of input-migrated volume (**Figure 12a**), similarity (**Figure 12b**), and similarity overlaid with TFL (**Figure 12c**). The similarity attribute effectively delineates the subsurface discontinuities. It also highlights the major and thick discontinuities in the survey area. The TFL result describes major faults in the center of the survey area as having a high likelihood of being larger and longer. Additionally, the TFL values in the southern area show the presence of relatively more fractures on a small scale. The subsurface features from the similarity and TFL attributes show good consistency, but the TFL attribute delineates minor discontinuities that are difficult to detect with the similarity attribute. Furthermore, the TFL attribute provides a more intuitive illustration of subsurface discontinuities and examination of probable fault images (**Figure 12d**).

6. Discussion and summary

A 3D-HRS survey was conducted to clearly reveal and map faults and fractures in a shallow-water region of Beppu Bay, Oita Prefecture, Japan. The high-frequency seismic source, densely distributed layout of seismic source and receiver points, precise positioning system, and data processing focused on noise and footprint suppression achieved high-resolution and high-precision imaging.

The 3D-HRS system, which is lightweight, easy to deploy and retrieve, and fits on a small vessel, realizes highly cost-effective data acquisition in shallow coastal waters, where data acquisition is difficult with conventional seismic survey methods using large vessels. Our results confirm that the 3D-HRS system can visualize subsurface structures down to depths of about 300 m below the seafloor surface at high resolution and that it is an effective tool for providing detailed information on complex systems of faults or fractures.

The 3D-HRS technique can be used to interpret the geological structure with a 3D seismic cube having various attributes and to understand the details of the faulting history based on the accumulation of fault displacements in the Holocene and upper Pleistocene deposits, which will greatly contribute to the evaluation of faulting activity in areas where trenching is not feasible. The evaluation of spatial characteristics of the subsurface geological structure and fault distribution using the 3D-HRS system is an extremely effective method for active fault investigations, including lateral strike-slip fault systems, and is considered to produce valuable data for regional earthquake disaster prevention.

Future work is expected to include more precise attribute and seismic facies analyses focusing on geological risks, such as fault fracture zones, shallow gas, submarine landslide deposits, and buried channel fill sediments. In addition, the importance of understanding the 3D subsurface structure with high precision is expected to increase in a wide range of fields, including geotechnical investigations for offshore wind power generation; site risk assessment of facilities; investigation of offshore resources such as oil and natural gas, methane hydrate, and seafloor minerals; and carbon dioxide capture and storage. The 3D-HRS system is expected to play an important role as a fundamental technology.

Acknowledgements

The 3D-HRS trial survey in Beppu Bay was conducted in collaboration with the Geothermal Research Facility of Kyoto University. We express our deepest gratitude to Professor Shinji Osawa and Professor Emeritus Keiji Takemura of Kyoto University for their support in conducting the survey.

Author details

Yousuke Teranishi[1], Takeshi Kozawa[1], Hitoshi Tsukahara[1], Fumitoshi Murakami[1], Yasuto Itoh[2*] and Shigekazu Kusumoto[3]

1 JGI, Inc., Tokyo, Japan

2 Osaka Metropolitan University, Osaka, Japan

3 Institute for Geothermal Sciences, Kyoto University, Oita, Japan

*Address all correspondence to: yasutokov@yahoo.co.jp

References

[1] Kawamura K, Nomura H. Off-shore wind power farms and marine geological risks. Journal of Japan Geotechnical Society. 2021;**69**(9):35-38

[2] Murakami F, Furuya M, Kochi E, Maruyama K, Hatakeyama K, Takeda N, et al. Development of the high-resolution three-dimensional seismic survey system for shallow water and the survey of active fault in the nearshore waters of the northern Suruga Bay using the system. Active Fault Research. 2016;**44**:29-40

[3] Ino S, Suda S, Kikuchi H, Okawa S, Abe S. Ultra-high-resolution 3D (UHR3D) seismic survey – application to Hinagu fault zone, Yatsushiro Sea, Kyushu, Japan – 2018. Butsuri-Tansa. 2018;**71**:33-42

[4] Suda S, Kozawa T, Murakami F, Sato N, Tsukahara H, Konno K. Trial survey of dual source UHR2D seismic system. In: Proceedings of the 143rd SEGJ Conference. Tokyo; 2020. pp. 131-134

[5] Takahashi K, Ikeda M, Sato T, Adachi K, Nishizaka N, Onishi K, et al. Distribution and activity of Median Tectonic Line in the Iyo-Nada Sea off northwestern Shikoku based on seismic survey. Active Fault Research. 2020;**53**:13-32

[6] Yusa Y, Takemura K, Kitaoka K, Kamiyama K, Horie S, Nakagawa I, et al. Subsurface structure of Beppu Bay (Kyushu, Japan) by seismic reflection and gravity survey. Zisin (Bulletin of Seismological Society of Japan). 1992;**45**:199-212

[7] MEXT (Ministry of Education, Culture, Sports, Science and Technology, Japan), Graduate School of Science at Kyoto University. Comprehensive Research and Survey for the Beppu-Haneyama Fault System (Oita Plain – Eastern Part of Yufuin Fault), Heisei 26-28 Fiscal Year Report. Kyoto: Graduate School of Science at Kyoto University; 2017. Available from: https://www.jishin.go.jp/database/project_report/beppu_haneyama/

[8] Itoh Y, Kusumoto S, Takemura K. Evolutionary process of Beppu Bay in Central Kyushu, Japan: A quantitative study of the basin-forming process controlled by plate convergence modes. Earth, Planets and Space. 2014;**66**:74. Available from: http://www.earth-planets-space.com/content/66/1/74

[9] Hale D. Methods to compute fault images, extract fault surfaces, and estimate fault throws from 3D seismic images. Geophysics. 2013;**78**:33-43

Chapter 5

Discussion

Yasuto Itoh, Akio Hara and Taiki Sawada

Abstract

The subsurface structure and sedimentary facies in the Beppu Bay basin were studied using 2D seismic and 3D high-resolution seismic (3D-HRS) surveys. A series of high-angle faults extend near the southern bay coast and are interpreted as active traces of the laterally moving Median Tectonic Line (MTL). A releasing bend of the MTL forms a pull-apart sag around the bay bottom accompanied by numerous normal fractures, whereas a compressional bulge is emerging on the southern side of the bay, suggesting complicated and transient stress states in the late Quaternary strata. Based on the general development history of the Hohi Volcanic Zone (HVZ), the lower, middle, and upper seismic horizons are assigned to 5–6 Ma (initial stage of the HVZ), 0.7 Ma, and 0.3 Ma, respectively. We identified three auxiliary reflectors within the upper part of the sediment pile and correlated them with oxygen isotope stages 15–11. These stratigraphic constraints are used to discuss paleoenvironments in the latest Pleistocene by means of an amplitude root mean square attribute analysis of the 3D-HRS data, which successfully delineated a buried river channel that was active during the last glacial period.

Keywords: Beppu Bay, Median Tectonic Line, seismic interpretation, seismic attribute, paleoenvironment

1. Introduction

As documented in Chapter 3, the latest technology of 3D high-resolution seismic (3D-HRS) surveying has successfully delineated fine structural features within the analyzed volume around the mouth of Beppu Bay (**Figure 1**) adjacent to the previous 2D seismic lines. In this chapter, we describe the deep fault architecture and stress–strain state around the bay in a wider scope using the conventional 2D data [1] and then discuss the possible transition of the pull-apart depocenter based on available age constraints of notable reflectors in the seismic profiles. Finally, the potential value of attribute analyses of the 3D-HRS dataset is presented from the viewpoint of paleoenvironmental interpretation.

2. Complicated stress distribution around Beppu Bay

Figure 2 shows a bird's-eye view of the subsurface structure of Beppu Bay [1, 2]. As depicted in profiles E and H, a series of high-angle faults having a "flower-like" appearance extend near the southern margin of the bay. Previous studies [2–4] interpreted that this fault zone is an active trace of the dextral Median Tectonic Line (MTL). Around the

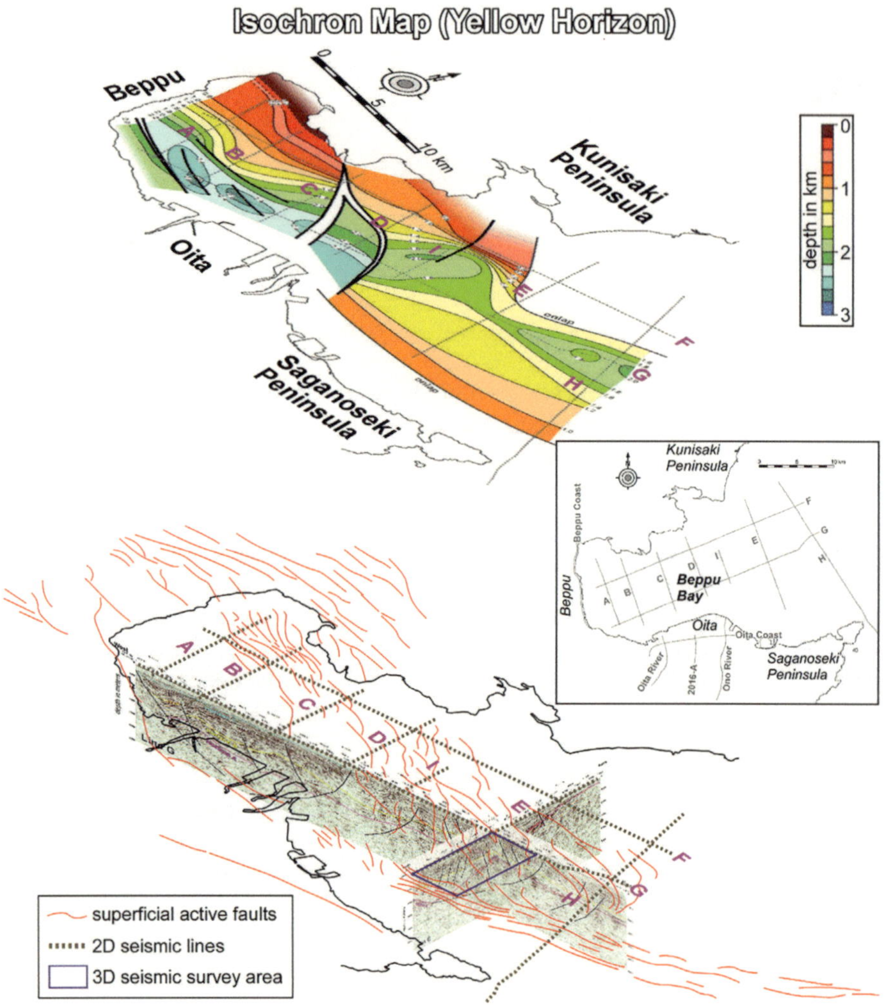

Figure 1.
Seismic survey index maps of the study area. The isochron map of the Pleistocene seismic horizon (upper) is after [1].

bay bottom (westernmost part), a releasing bend of the MTL has formed a pull-apart sag, the outline of which can be traced along numerous normal fractures in the shallows (red lines in **Figure 1**). Although such extensional features are suggestive of a prevalent tensile stress field [5]; Itoh and Inaba [2] argued that a compressional bulge is growing under the southern bay coast from the structural interpretation of recent obvious inversion on a seismic section (**Figure 2b**). Thus, Beppu Bay is characterized by a complicated and transient stress state during the late Quaternary. Temporal constraints on the architectural buildup of the study area are discussed in the next section.

3. Age assignment of seismic horizons

The Quaternary system exposed around the bay is dominantly composed of volcaniclastic material derived from a number of active volcanoes. Their varied lithofacies

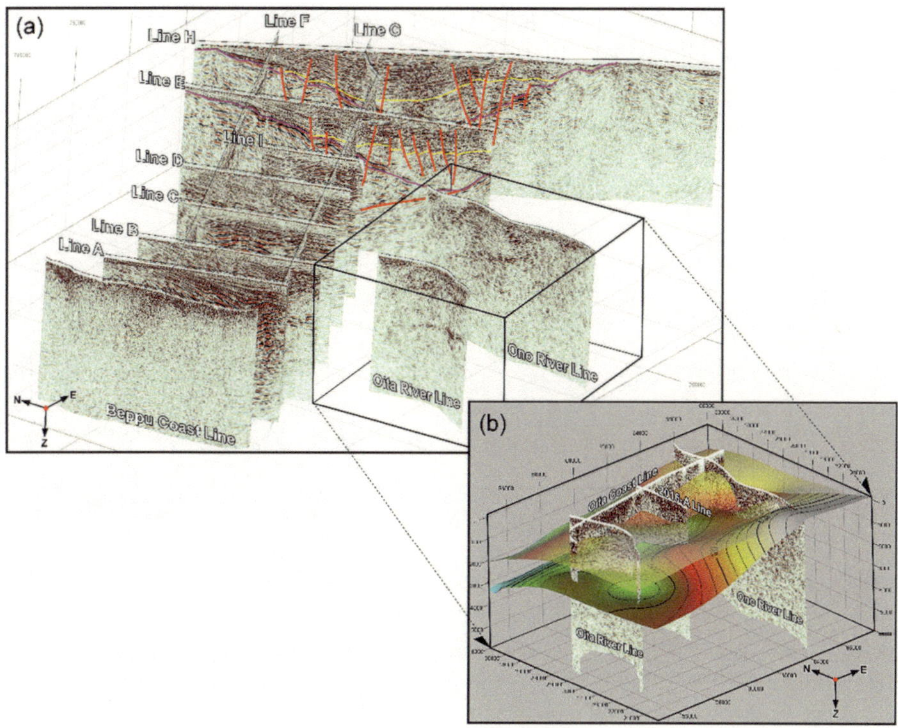

Figure 2.
Bird's-eye views of the 2D seismic data (modified from [1]) around Beppu Bay.

and absence of index fossils hinder precise stratigraphic correlation. Moreover, the ages of significant seismic horizons have not been established for the sediments filling the Beppu Bay basin due to the lack of borehole data. To overcome such difficulties, we assign numerical ages to conspicuous horizons based on major geologic events recorded in terrigenous sediments. **Figure 3** presents our stratigraphic model. Based on the general development history of the Hohi Volcanic Zone (HVZ) tectonic graben [4], the bottom of the basin fill (Horizon A) is assumed to be 5–6 Ma. Another ubiquitous horizon (B) in the middle of the sediment pile is correlated with an unconformity at the basal part of the Oita Group (ca. 0.7 Ma [6]). The uppermost Horizon C, traceable within the western half of the bay, coincides with a drastic change in seismic facies and is interpreted as the cessation of clastic supply from the Hiji volcano on the northern coast (ca. 0.3 Ma [7]). Major seismic datum levels are thus assigned to volcano-tectonic events.

Other semicontinuous reflectors are probably related to paleoenvironmental changes. Beppu Bay is an estuary up to of 70 m deep, and it has inevitably experienced marine/nonmarine transition according to eustatic changes in sea level. Previous stratigraphic studies in similar environments [8, 9] showed that traceable seismic reflectors are linked with basal horizons of marine clays, which are clearly recognized by sharp lithologic changes, and have straightforward correspondence to the oxygen isotope record during rapid marine transgression. Based on these chronological constraints, the evolution and migration of pull-apart basins in the bay area is reconstructed in the next section.

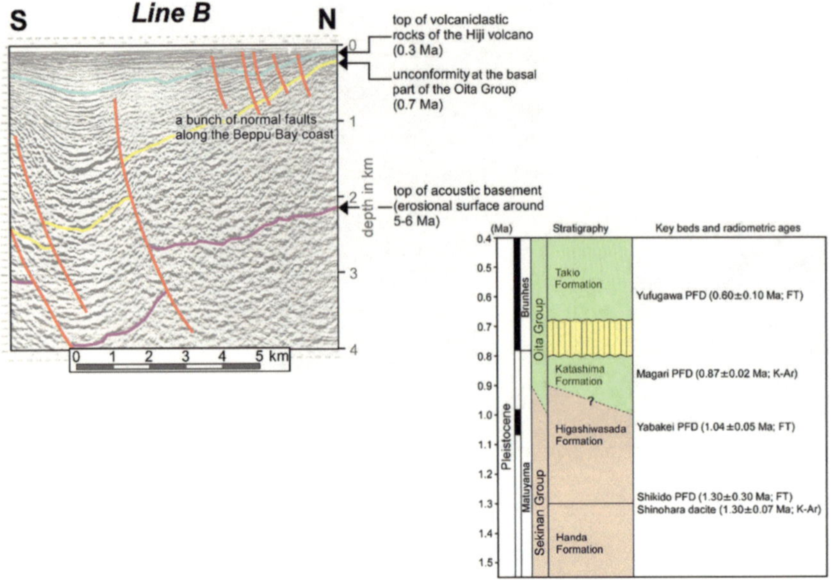

Figure 3.
Chronological framework of 2D seismic horizons and stratigraphic summary of the Quaternary system exposed around Beppu Bay. See the text for data sources.

4. Depocenter migration within Beppu Bay

Figure 4 presents a record of a 2D seismic line running from the bottom (west) to the mouth (east) of the bay. The major horizons A, B, and C described above are traced on the profile and delineate remarkable fault-bounded depressions. The two depocenters accompanied by extensional features are clearly seen. They are aligned along the MTL active fault system (**Figure 2**) and interpreted as a series of pull-apart basins. It is noteworthy that the displacement growth on the faults bounding the western depocenter is indicative of ongoing subsidence, whereas the eastern one seems to have been dormant in recent periods. This phenomenon is probably related to depocenter migration provoked by terminal propagation of the MTL, as is often the case with large transcurrent faults (e.g., [11, 12]).

Based on their numerical ages, Horizons B and C are assigned to global oxygen isotope stages 17 and 9, respectively (after [10]). Between these controlling datum levels are three auxiliary reflectors (see **Figure 4**). As explained in the previous section, they originate from eustatic marine transgression and are correlated with marine isotope stages (MISs) 15–11 (odd numbers) in ascending order, considering the stratigraphic framework. Thus, an evolutionary overview of Beppu Bay, described by Itoh and Inaba [2], is obtained from the 2D seismic interpretation.

5. Insight into paleoenvironments

5.1 3D seismic interpretation

The 3D-HRS survey data were acquired by Kyoto University and JGI, Inc., within Beppu Bay. The survey area location, ranging 3 km by 6 km in the bay and located

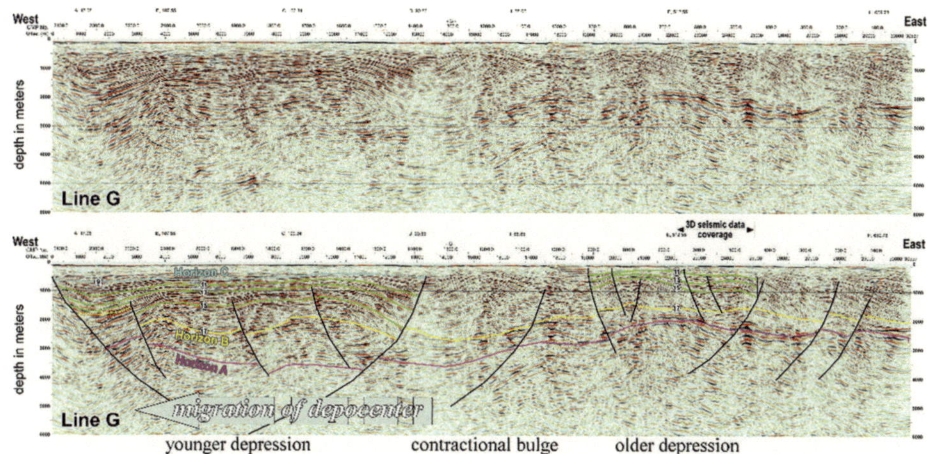

Figure 4.
E-W seismic profile (G; see **Figures 1** *and* **2** *for location) in Beppu Bay. Raw section (top) and interpreted section (bottom) are shown after depth conversion. See* **Figure 3** *for numerical ages of the controlling horizons. Annotation numbers on the major and minor (green) horizons are oxygen isotope stages after Lisiecki and Raymo [10].*

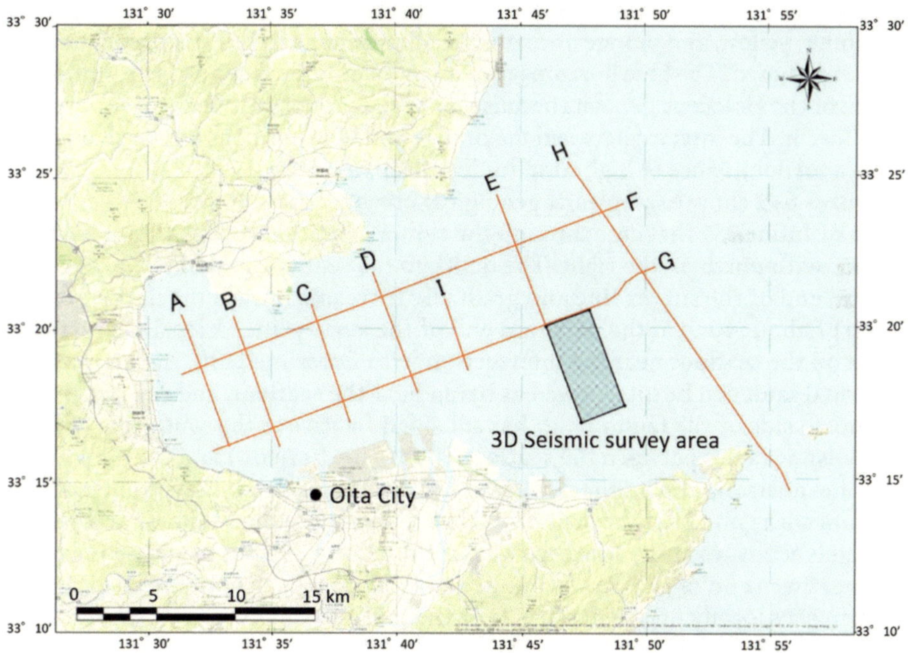

Figure 5.
Location map of the 3D seismic survey area.

15 km northeast offshore of Oita City (Oita Prefecture), is shown in **Figure 5**. The northern and western edges of the survey area are in contact with the existing 2D seismic survey lines G and E, respectively.

Geological interpretation work was carried out by combining the geological structure interpretation obtained by tracing the reflecting surface and the attribute

Horizon	Age	Remarks
Light Blue	Resent	Sea floor
Blue	Holocene	Base of Alluvial
Orange	Pleistocene	Base of High amplitude unit
Yellow		Reflective surface with good continuity. Age unknown.
Pink		
Green		
Light Green		MIS-11 (Estimated)
Magenta		MIS-13 (Estimated)

Table 1.
List of interpreted horizons used for geological interpretation.

analysis obtained by applying the amplitude root mean square (RMS). For the geological interpretation, we selected eight reflective surfaces that can be traced continuously within the survey area. **Table 1** shows the names and characteristics of each horizon. The lowest magenta horizon is a reflective surface that can be compared with the MIS-13 horizon (ca. 0.49 Ma) estimated from the 2D cross section shown in **Figure 4**. Also, the light green horizon can be compared to MIS-11 (ca. 0.41 Ma). The green, pink, yellow, and orange horizons are all presumed to be Pleistocene, but their ages are unknown. The blue horizon is an unconformity surface and is presumed to be the base of the Holocene deposits because modern seafloor sediments have accumulated above it. The interval between the orange and blue horizons can be characterized by a predominance of high-amplitude reflective surfaces.

Figures 6–9 show examples of geological cross sections. **Figure 6** is a cross section of Inline 251 that cuts through the center of the survey area from north to south, with north on the right. The depth to the seafloor is about 25 m at the southern end of the survey line and gradually increases toward the north, reaching a depth of about 40 m at the northern end of the survey line. A local depression is present on the seafloor near the intersection with Crossline 1100. At this position, the normal fault can be interpreted as being near the seafloor, and the hanging wall (north side of the fault plane) has subsided relative to the south side of the fault. Seismic facies between the seafloor (light blue horizon) and the alluvium base (blue horizon) can be characterized by a predominance of low-amplitude reflectors with good lateral continuity. The alluvium becomes thinner and almost completely vanishes at the southern end of the survey line but reaches a maximum at the northern end of the survey line at about 50 m. In addition, the layer thickness increases locally on the north side of the fault at Crossline 1100. The alluvium base is an unconformity surface, and the underlying reflective surface is cut by the alluvium base. Underneath the alluvium base strata, the geology is dominated by discontinuous, strong-amplitude reflective surfaces. This geologic unit is composed of multiple strongly reflective surfaces, and the base of this unit is traced as a high-amplitude base horizon (orange horizon).

The thickness of the geological body is about 30 m at the north end of the survey line, but it is denuded by the alluvium base near Crossline 700 and disappears. The age of this geological unit is estimated to be Pleistocene because it is directly denuded by the alluvium base. The strongly reflective surfaces below the orange horizon are identified and tracked as yellow, pink, green, light green, and magenta from the top.

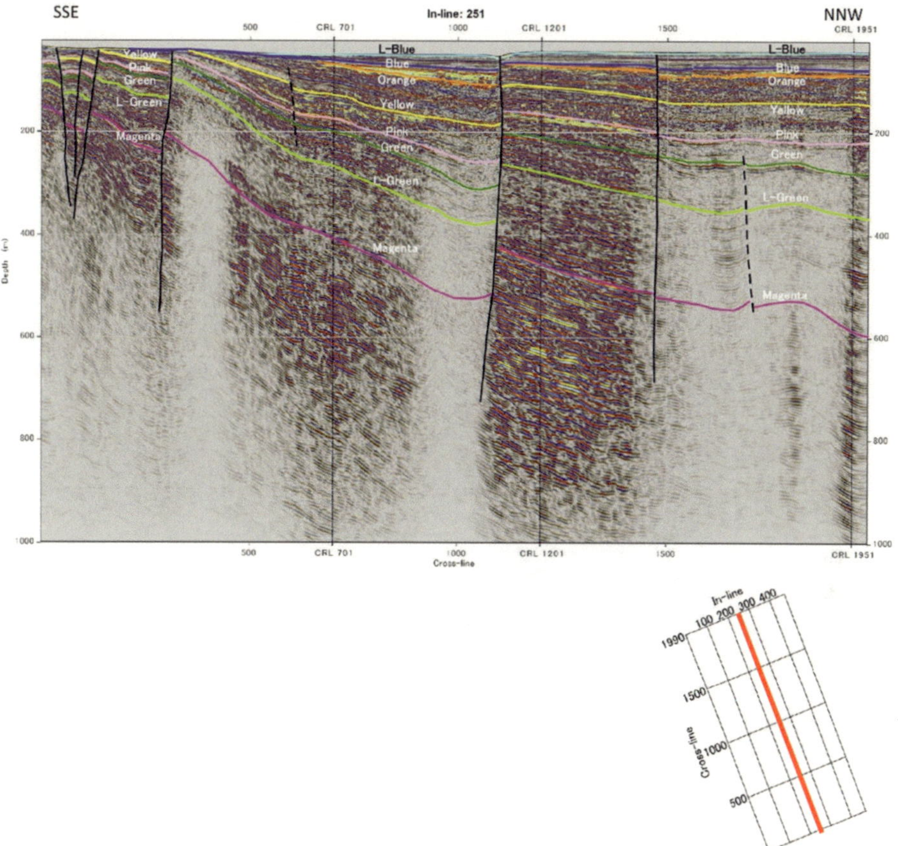

Figure 6.
Geological interpretation section of inline 251. Thick black lines are faults. Light green and magenta horizons are assigned to the MIS-11 and MIS-13 horizons [10], respectively.

It is difficult to determine the geological age because no boreholes have been drilled into the research area. Between the orange and yellow horizons, reflective surfaces with relatively low amplitude and poor continuity predominate. The layer thickness of this section is about 60 m at the north end of the survey line, but it tends to become thinner toward the south. Between the yellow and pink horizons, a reflective surface with a strong amplitude but poor continuity dominates. The layer thickness is about 70 m at the north end of the survey line, and it decreases toward the south to about 40 m at the south end of the survey line. The green horizon can be characterized by a continuous strong-amplitude reflective surface. The reflective surfaces between the pink horizon and the green horizon are dominated by relatively low-amplitude reflective surfaces with poor continuity. Reflective surfaces in the section from the green horizon to the magenta horizon can be characterized by weaker amplitudes and more prominent reflective surfaces with good continuity compared to that of the upper sections. The layer thickness between the green horizon and the light green horizon is about 70 m at the north end of the survey line and about 20 m at the south end of the survey line. The layer thickness of the light green and magenta horizons is about 200 m at the north end of the survey line and about 50 m at the south end of the survey line.

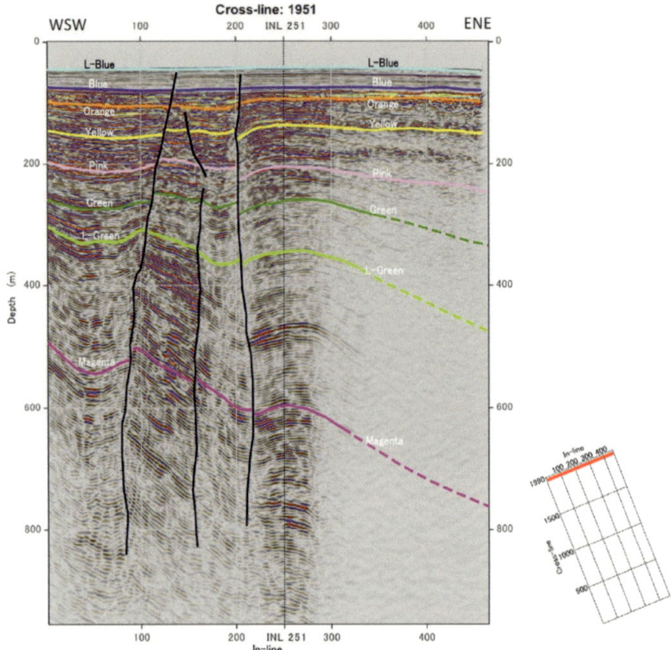

Figure 7.
Geological interpretation section of crossline 1951. Thick black lines are faults. Light green and magenta horizons are assigned to the MIS-11 and MIS-13 horizons [10], respectively.

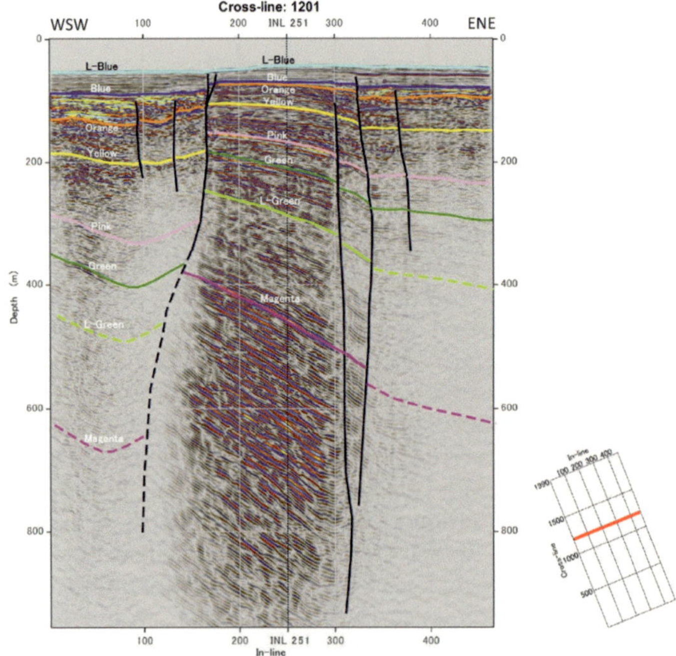

Figure 8.
Geological interpretation section of crossline 1201. Thick black lines are faults. Light green and magenta horizons are assigned to the MIS-11 and MIS-13 horizons [10], respectively.

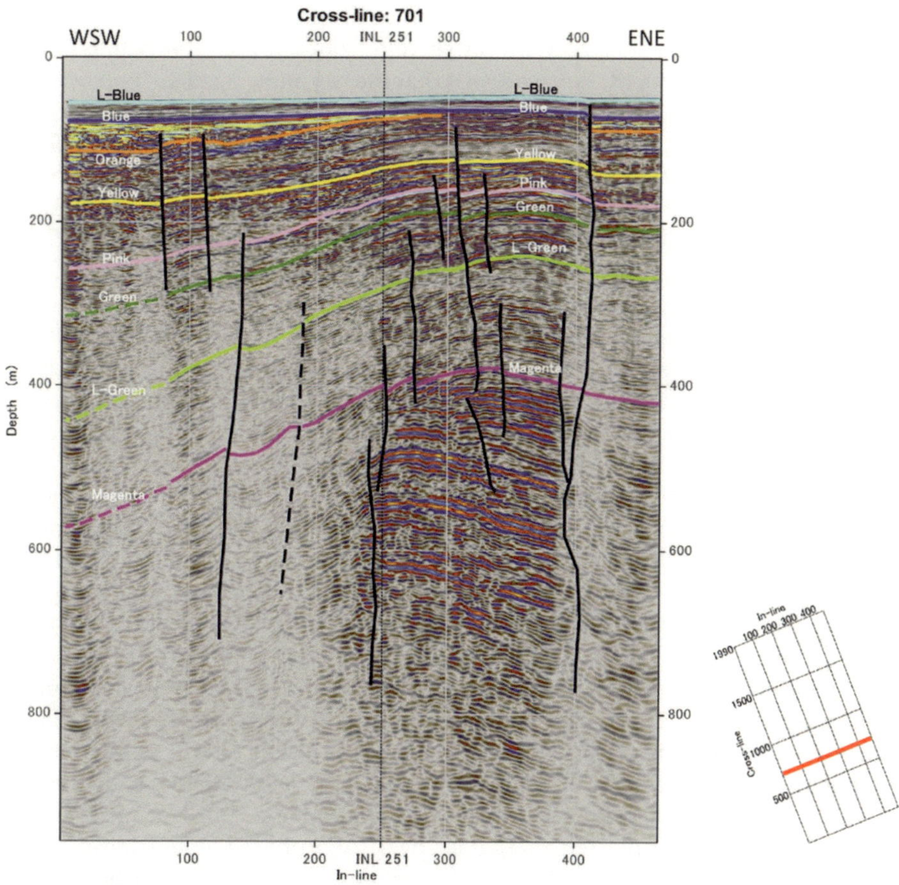

Figure 9.
Geological interpretation section of crossline 701. Thick black lines are faults. Light green and magenta horizons are assigned to the MIS-11 and MIS-13 horizons [10], respectively.

Figure 7 shows the geological interpretation of Crossline 1951. This cross section is located at the northern end of the 3D seismic survey and is close to the existing 2D survey line G [13], so it can be compared. In this chapter, we compare the geological ages of the reflection surfaces on line G with the high sea level period from MIS-17 to MIS-9 (see **Figure 4**). Regarding the 3D data acquisition range, MIS-13 and MIS-11, which are distributed at a depth of less than 1000 m, are compared with the magenta horizon and light green horizon, respectively.

The thickness distribution of strata in Crossline 1951 tends to increase toward the northeast side (right side). Three high-dip faults can be identified near the center of the cross section. A synclinal structure is formed on the reflecting surface that is in contact with the westernmost fault. In addition, an anticline structure can be identified on the northeast side of the easternmost fault. The reflecting surface in the section between these two faults (generally the section between Inlines 100 and 200) dips to the east, and the dip tends to increase with depth. In addition, at the eastern edge of this section (e.g., Inline 200, depth 600 m), a local synclinal structure can be discerned near the contact with the fault. The azimuth and dip of the fault plane are not constant, and a locally bulging shape can be seen (e.g., Inline 100, depth 400 m).

Figure 8 shows the geological interpretation of Crossline 1201. The geological block in the center of the cross section, the section from Inlines 170 to 320, is bounded by faults and slopes eastward. It can be seen that the dip angle tends to increase as the depth increases. The geological unit from the western edge of the cross section to Inline 170 has subsided greatly along the fault plane, and the alluvium is thicker than the central geological block, reaching approximately 40 m. It is accompanied by multiple small-scale normal faults, and a flower-like structure can be seen. In addition, for the geological unit east of Inline 320, we can interpret the geological "flower" structure as comprising multiple normal faults. At Crossline 1201, both sides of the central geological block subside due to normal faults.

Figure 9 shows the result of interpretation of Crossline 701. At Crossline 701, the west-dipping geological structure is dominant. Many small-scale, high-dip faults are formed near Inline 300, and individual blocks bounded by faults tilt to the west, forming a sawtooth-like structure. This situation has been interpreted in the pink and green horizons. The fault located at Inline 400 is a normal fault, and the adjacent eastern geological unit is subsiding. In the section from Crosslines 250 to 400, the dip of the reflecting surface to the east with the anticline axis near Inline 270 below the magenta horizon is dominant, and above that, the position of the anticline axis is eastward. Around the yellow horizon, the west-dipping structure around Inline 370 becomes dominant.

Geological interpretation work was conducted using 3D high-resolution earthquake exploration data. **Figure 10** shows the depth structural diagram of the created seabed (light blue), the offshore layer base (blue), and the update (yellow and green). The inset shown at the upper left is the depth structure of the seabed surface. The water depth is about 30 m at the southern end near the coast, and the depth increases farther offshore, becoming about 40 m around Crossline 400. There is a distinctive trend in the depth distribution on the north side of Crossline 400. The most distinctive seabed terrain is cliff-like terrain that is linearly distributed on the line connecting Inline 1, Crossline 1501 and Inline 451, Crossline 801. The southwest side of the cliff is sinking, forming the deepest water area that exceeds a depth of 50 m. Another topographical characteristic is a ridge-like terrain extending northwest along Inline 351. The water is deeper toward both sides (northwest and southeast) on the ridge.

One of the characteristics of the depth distribution in each layer below the sea floor is two linear geological structures caused by the fault. One is on a line connecting Inline 1, Crossline 1401 and Inline 451, Crossline 901, and the other starts around Inline 200, Crossline 1990; passing Inline 351, Crossline 801; and reaching near Inline 451, Crossline 601. Here, the former is called Fault 1, and the latter is called Fault 2. For Fault 1, the block southwest of the fault is subsiding. Regarding Fault 2, the northeast side of the fault is subsiding. Judging from the depth contour of the geological block sandwiched between Faults 1 and 2, an anticline extending along the fault is formed at the southwestern end of the block, and it seems to dip toward the northeast. The geological block on the southwest side of Fault 1 is sloping toward the northwest and is deepest near the fault. It is possible that this geological block forms a half-graben-like structure. In contrast, the block on the northeast side of the Fault 2 is tilted in the northeast direction.

5.2 Attributes of 3D seismic data

The RMS of the amplitude was applied to the 3D seismic survey data acquired in Beppu Bay to perform a survey attribute analysis. The RMS attribute was calculated as a high-amplitude zone in the section of the window containing the trace if the

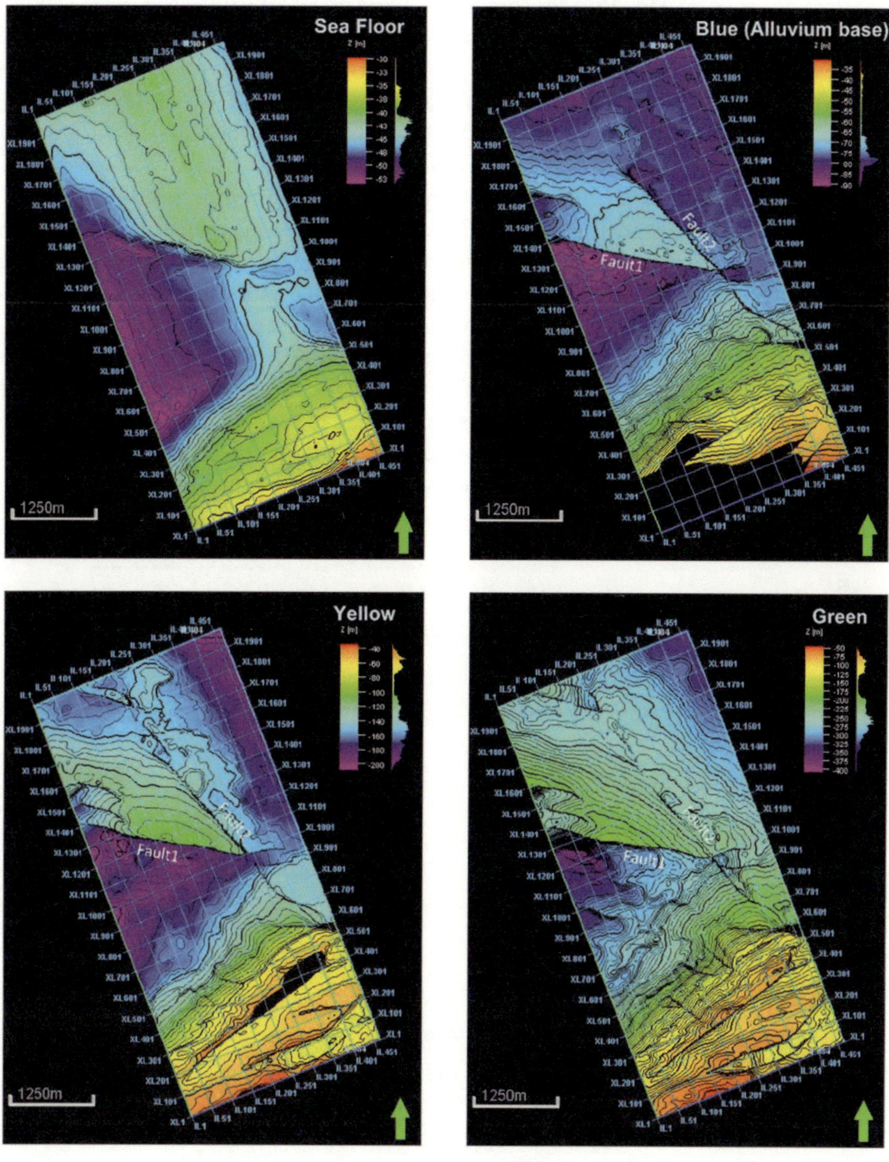

Figure 10.
Geological structure map of the sea floor, alluvium base, and Pleistocene deposits (yellow horizon and green horizon).

window section contains a trace sample with a large amplitude. The value was calculated for one trace of the seismic data, while that trace was moved over another trace by setting an arbitrary interval. The RMS amplitude attribute was calculated for the RMS of the amplitude value included in the seismic survey data. As an example of the reflection surface formed by a high amplitude, it was possible to identify a geological layer boundary where the difference between the impedance of the geological layer (product of the acoustic wave velocity and the density) increases. In the field of petroleum exploration, the RMS magnitude attribute is applied, and many cases of sandstone reservoirs with large impedances are used for the examination of their distributed areas (e.g., [14, 15]). In this study, Landmark Solution's SeisSpace®/ProMAX® was used for attribute analysis. The window size for calculating the RMS attribute was 11 samples (5 m).

Figure 11 shows the analysis results for amplitude RMS. The top-left image is a slice section of the amplitude RMS at a depth of 87 m, and the top-right image is the amplitude RMS at a depth of 76 m. At 87 m, the pattern of a meandering river is continuously observed from Inline 1, Crossline 600 to the northeast direction, in the green to red area showing high-amplitude RMS, and the blue area showing low-amplitude RMS can be tracked. The 87-m level is located in the depth interval from the orange horizon to the alluvium base (blue horizon) where the high-amplitude reflector is dominant. Therefore, it is estimated that the amplitude RMS also increases in this section. In contrast, the river sediment RMS has a smaller amplitude than that of the surrounding strata, making it possible to extract the pattern of the river channel. The top-right image in the figure is the amplitude RMS at a depth of 76 m. In this slice section, we were able to capture an amplitude RMS pattern indicative of the meandering river within an area where such a pattern

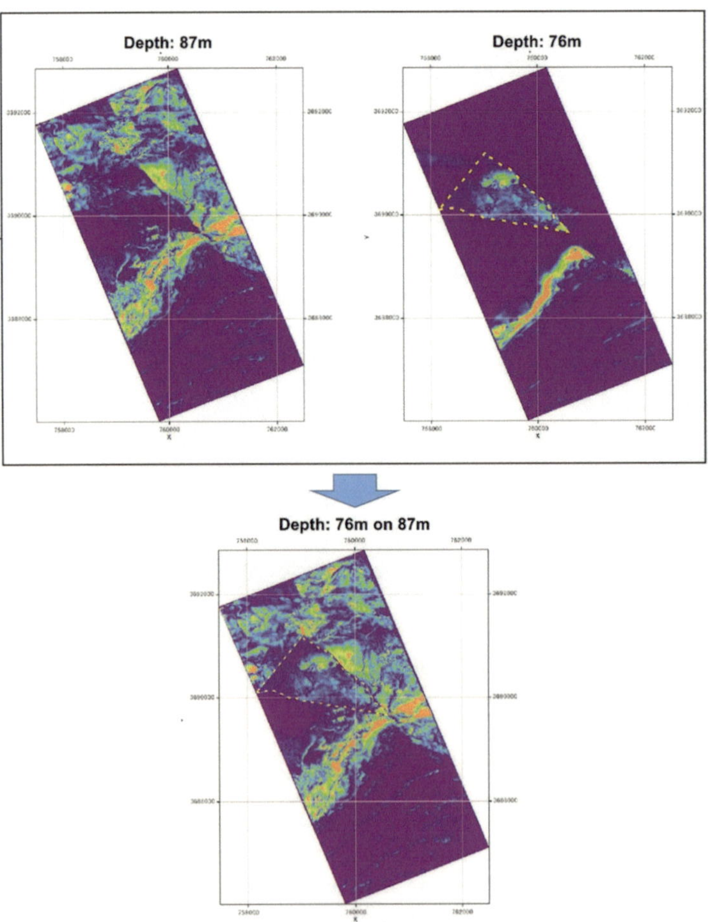

Figure 11.
Mosaic map of the old channel. Top left: Depth at 87 m. top right: Depth at 76 m. Bottom inset shows a combined map.

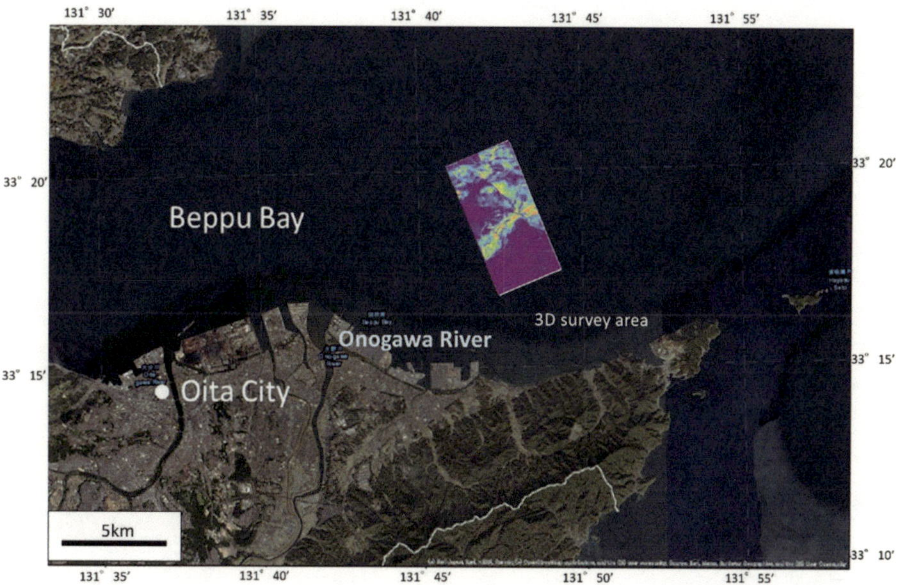

Figure 12.
Comparison between the position of the old channel extracted from the amplitude RMS analysis result and the current river position.

was missing in the slice section at 87 m. In the lower panel of **Figure 11**, which is a mosaic of amplitude RMS slices at both depths, the continuous meandering river pattern can be discerned.

Figure 12 shows the location of the channel extracted from the amplitude RMS attribute analysis in a satellite landscape. The river channel extracted from the amplitude RMS attribute flows from the southwest to the northeast near the center of the 3D seismic survey area. Because the river is divided into multiple channels, it is estimated that the river flowed through a floodplain and its channel changed several times. Based on this river channel distribution and the spread of the floodplain, it is highly likely that the channel will continue to the vicinity of the present-day Onogawa River if it is traced upstream. The Onogawa River is a river that flows northeastward through the eastern half of the Oita Plain. The 3D survey area is located on the extension of the current Onogawa River. There are no large rivers in the eastern Oita Plain. Judging from the extent of the present-day delta for the Oita Plain, the Onogawa River is considered to be the most likely river to have reached the 3D survey area during low sea level periods.

5.3 Geological structure that regulates the position of the old channel

Figure 13 shows a distribution diagram of the old channel in the slice of the geological interpretation section of Crossline 701 and the amplitude RMS with a depth of 87 m. The two yellow arrows shown at the western end of the geological interpretation section indicate the position where the old channel intersects this cross section. In this cross-sectional view, local undulations associated with faults are formed near the channel. The channel is located in a local concave area (near Inline 115), which is locally high (near Inline 90), and a high-rising edge. This suggests that the old channel of this era may have been controlled by the local uplift associated with the fault.

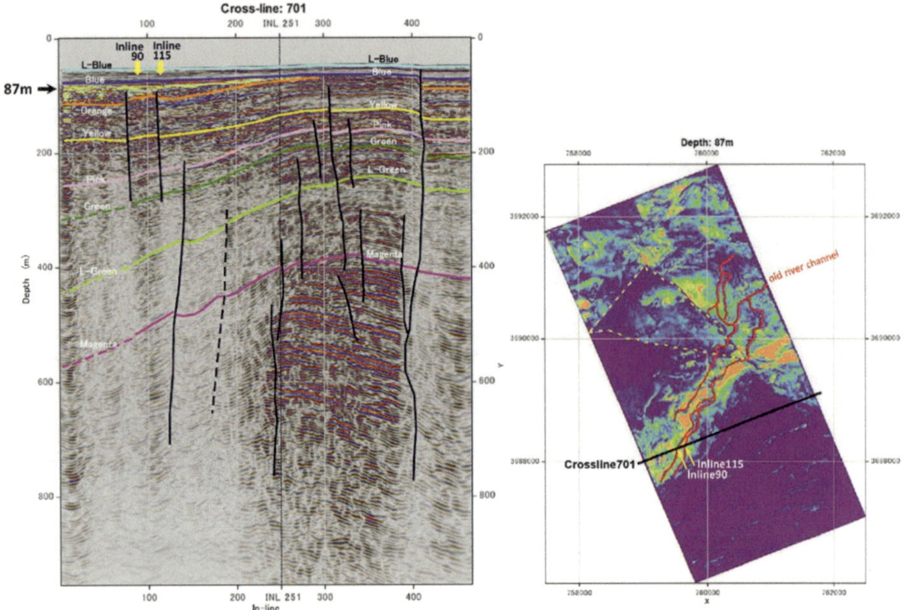

Figure 13.
Characteristics of the geological structure where the channel was formed, as seen in the seismic section.

Figure 14 shows the old channel flow path extracted from the results of the attribute analysis on the depth structure map of the orange horizon, where the channel is extracted. The channel is displayed as a red solid line. In this figure, the old river channel tends to have been formed in concave terrain and is consistent with the location of the old river path interpreted from the seismic section. An unconformity exists between the strata where the old channel was extracted and the alluvium, so it is difficult to identify the age of the sediments containing the old channel. However, considering that the depth difference from the basement of the alluvium is about 10 m, it is possible that the channel was formed during the late Pleistocene (near Holocene).

The bottom left of **Figure 14** shows the old river channel (solid red line) extracted as a result of attribute analysis superimposed on the depth map of the base of Holocene deposits (alluvium). In this figure, in the section from Inline 1 to Inline 350, the azimuth of the base of the alluvium is sloping northwest, while the old channel is diagonal. For this reason, it is difficult to assume that the river flowed in the northeast direction here. The bottom-right inset in **Figure 14** shows the old river channel superimposed on the current seafloor topography. In the section from Inline 1 to Inline 350, the channel is located on the slope of an up-dip. This suggests that the Onogawa River cannot flow northeastward through this section. Considering this series of trends, it is possible that the subsidence of the geological block located in the southwest direction progressed in this vicinity after the age when the old channel was formed. In the geological interpretation section shown in **Figure 13**, the anticline axis of the magenta horizon is located near Inline 300, but above it, the position seems to be gradually shifting to the east. This tendency seems to be consistent with the estimation that the Onogawa River had difficulty flowing northeastward during the geological age above the orange horizon.

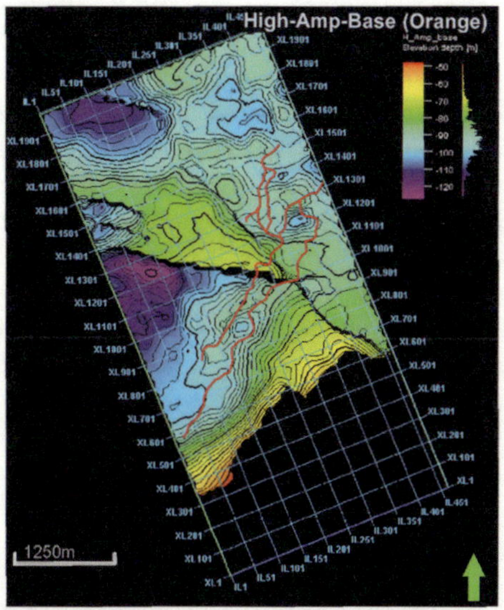

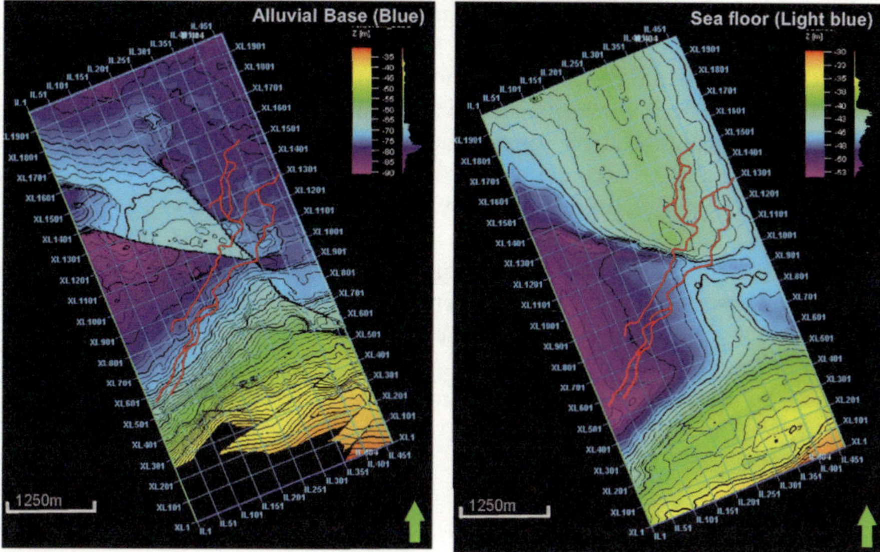

Figure 14.
Comparison between the terrain of the era when the old river channel was formed and the terrain of the subsequent geologic periods.

6. Conclusion

In this chapter, we considered the growth process of the pull-apart basins in Beppu Bay, which are thought to have formed with right-lateral movements on the MTL based on the existing 2D seismic survey data. In the paleoenvironmental discussion, the latest 3D high-precision seismic survey data elucidated the following points.

1. As a result of interpreting the high-resolution 3D seismic profile obtained in the southern region of Beppu Bay, the stratigraphic horizon of an unconformable

surface was assigned to the Holocene base, and lower stratigraphic horizons were extracted for comparison with MIS-13 and MIS-11.

2. Two faults (Faults 1 and 2) were detected near the center of the northern half of the 3D seismic survey data. These faults intersect at an acute angle. The "inverted triangular" geologic block between the faults is tilting to the northeast. In addition, a local anticline structure is formed along Fault 1 at the southwestern end of this block.

3. The geologic block adjacent to Fault 2 at the eastern end of the 3D seismic survey area is subsiding against the inverted triangular geologic block and dips to the northeast. In contrast, the block adjacent to the west side of Fault 1 is also subsiding against the inverted triangular geologic block and dips to the northwest. The latter geologic block forms a half-graben structure with Fault 1 as the main fault.

4. As a result of amplitude RMS attribute analysis on the 3D seismic data, we succeeded in extracting an old river channel from a geological unit underlying the base of the Holocene. The flow direction of this old river channel seems to northwest to northeast in the 3D seismic data area.

5. The depths of the old river channel are in the range of around 86 m from the sea surface. However, it was also detected at a depth of ca. 76 m at the southern end of the geological block between two faults (i.e., Faults 1 and 2). The latter portion is about 10 m shallower than the surrounding areas, implying pop-up of a faulted block after the periods of channel growth.

6. The estimated flow direction of the old river channel is northeastward, but the Holocene base and modern seafloor tend to incline northwestward, so it is difficult to imagine a river flowing northeastward at this location. This fact suggests that a tectonic event provoking northwestward subsidence in the geological block west of Fault 1 may have been underway since the period of the ancient river flow.

Author details

Yasuto Itoh[1*], Akio Hara[2] and Taiki Sawada[2]

1 Osaka Metropolitan University, Osaka, Japan

2 JGI, Inc., Tokyo, Japan

*Address all correspondence to: yasutokov@yahoo.co.jp

References

[1] MEXT (Ministry of Education, Culture, Sports, Science and Technology, Japan), Graduate School of Science at Kyoto University. Comprehensive Research and Survey for the Beppu-Haneyama Fault System (Oita Plain – Eastern Part of Yufuin Fault), Heisei 26-28 Fiscal Year Report. Kyoto: Graduate School of Science at Kyoto University; 2017. Available from: https://www.jishin.go.jp/database/project_report/beppu_haneyama/

[2] Itoh Y, Inaba M. Fault-Controlled Processes of Basin Evolution: A Case on a Longstanding Tectonic Line. New York: Nova Science Publishers, Inc.; 2019. p. 81

[3] Itoh Y, Kusumoto S, Takemura K. Evolutionary process of Beppu Bay in Central Kyushu, Japan: A quantitative study of the basin-forming process controlled by plate convergence modes. Earth, Planets and Space. 2014;**66**:74. Available from: http://www.earth-planets-space.com/content/66/1/74

[4] Itoh Y, Takemura K, Kamata H. History of basin formation and tectonic evolution at the termination of a large transcurrent fault system: Deformation mode of Central Kyushu, Japan. Tectonophysics. 1998;**284**:135-150

[5] Shimazaki K, Nakata T, Chida N, Miyatake T, Okamura M, Shiragami H, et al. A preliminary report on the drilling project of submarine active faults beneath Beppu Bay, Southwest Japan, for longterm earthquake prediction. Active Fault Research. 1986;**2**:83-88

[6] Yoshioka T, Hoshizumi H, Miyazaki K. Geology of the Oita District with Geological Sheet Map at 1:50,000. Tsukuba: Geological Survey of Japan, AIST; 1997. p. 65

[7] Ishizuka Y, Mizuno K, Matsuura H, Hoshizumi H. Geology of the Bungo-Kitsuki District with Geological Sheet Map at 1:50,000. Tsukuba: Geological Survey of Japan, AIST; 2005. p. 83

[8] Itoh Y, Takemura K, Ishiyama T, Tanaka Y, Iwaki H. Basin formation at a contractional bend of a large transcurrent fault: Plio-Pleistocene subsidence of the Kobe and northern Osaka Basins, Japan. Tectonophysics. 2000;**321**:327-341

[9] Itoh Y, Takemura K. Three-Dimensional Architecture and Paleoenvironments of Osaka Bay - An Integrated Seismic Study on the Evolutionary Processes of a Tectonic Basin. Singapore: Springer Nature Singapore Pte Ltd.; 2019. p. 119

[10] Lisiecki LE, Raymo ME. A Pliocene-Pleistocene stack of 57 globally distributed benthic $\delta^{18}O$ records. Paleoceanography. 2005;**20**:PA1003. DOI: 10.1029/2004PA001071

[11] Noda A. Strike-slip basin - its configuration and sedimentary facies. In: Itoh Y, editor. Mechanism of Sedimentary Basin Formation – Multidisciplinary Approach on Active Plate Margins. Rijeka: IntechOpen; 2013. DOI: 10.5772/56593

[12] Noda A, Toshimitsu S. Backward stacking of submarine channel-fan successions controlled by strike-slip faulting: The Izumi Group (Cretaceous), Southwest Japan. Lithosphere. 2009;**1**(1):41-59. DOI: 10.1130/L19.1

[13] Yusa Y, Takemura K, Kitaoka K, Kamiyama K, Horie S, Nakagawa I, et al. Subsurface structure of Beppu Bay (Kyushu, Japan) by seismic reflection

and gravity survey. Zisin (Bulletin of Seismological Society of Japan). 1992;**45**:199-212

[14] Rao PH, Hansa GL, Savanur S, Mangal S, Ramegowda B, Shanker L, et al. Detection of thin sandstone reservoirs using multi attribute analysis and spectral decomposition on post stack 3D seismic data, north Sarbhan oil field, south Cambay basin, Gujarat, India. In: Proceedings of 5th Conference & Exposition on Petroleum Geophysics, Hyderabad-2004. Hyderabad, India: Society of Petroleum Geophysicists; 2004. pp. 752-759

[15] Azeem T, Wang Y, Khalid P, Liu X, Fang Y, Cheng L. An application of seismic attributes analysis for mapping of gas bearing sand zones in the Sawan gas field, Pakistan. Acta Geodaetica et Geophysica. 2016;**51**:723-744. DOI: 10.1007/s40328-015-0155-z

Chapter 6

Epilogue: Scientific Findings and Perspective for Future Work

Yasuto Itoh

The multidisciplinary research described in this book has elucidated the entire evolutionary process of a tectonic basin, Beppu Bay, located at the termination of a large strike-slip fault bisecting the southwestern Japan arc.

1. The Median Tectonic Line (MTL) active fault system in southwest Japan is a typical arc-bisecting fault driven by oblique subduction of the Philippine Sea Plate through late Quaternary period. At the western termination of the MTL active fault system, the Beppu Bay basin in central Kyushu Island has been developing since the Pliocene as the eastern part of a volcano-tectonic depression of the Hohi Volcanic Zone (HVZ), of which the outline and dimensions (more than 5000 km^3) were delineated by means of gravimetric analyses.

2. The HVZ is a box-shaped graben buried by Plio-Pleistocene volcaniclastic materials. Although its genesis is often attributed to N-S breakup of the continental crust of Kyushu, recent horizontal movements detected by GNSS-based analysis have contradicted this hypothesis. Instead, the HVZ is related to the extrusion tectonics provoked by westward indentation of the forearc sliver of southwest Japan according to intermittent dextral slips on the MTL, which has been active as a right-lateral fault system under the control of west-northwestward convergence of the Philippine Sea Plate since the late Pleistocene. The right-stepping configuration of the dextral fault resulted in the emergence of an active pull-apart basin around Beppu Bay.

3. The basement architecture of the HVZ, including Beppu Bay, was reproduced using dislocation modeling. Development of the HVZ is divided into three stages: formation of a half-graben (Stage I, Pliocene), formation of the initial pull-apart basins due to activation of right-lateral faults (Stage II, early Quaternary), and growth of the pull-apart basins due to changes in the active area of the right-lateral fault (Stage III, middle to late Quaternary). During the course of optimization, a two-layer model of density contrast (basement and sedimentary layer) was introduced around the study area. The calculated gravity anomaly caused by the basement topography matched well with observed anomalies. Thus, the geological evolutionary model of the HVZ is endorsed from a geophysical viewpoint.

4. High-resolution 3D seismic (3D-HRS) survey data were acquired in Beppu Bay. The survey area was a rectangle of 6 km by 3 km (18 km^2). The compact 3D-HRS data acquisition system was developed for detailed understanding of geologic

IntechOpen

structures of the shallow subsurface, and it comprises short streamer cables, small high-frequency seismic sources, and onboard equipment. As the survey area is congested with many ships at all times, data acquisition was conducted using a small vessel equipped with approximately 100-m-long streamer cables. During the acquisition of seismic data, the streamer spread was automatically optimized by the steering devices mounted on the tail floats of the cables, to ensure that the towing interval of the streamer cables remained in a specified range. After a series of data processing steps (pre-stack noise attenuation, multiples removal, velocity analysis, and footprint removal), seismic attribute calculations (similarity and thinned fault likelihood) were conducted to provide a quantitative basis for the interpretation of geologic features. The 3D-HRS method has much higher resolution than existing 2D seismic sections. In particular, it effectively extracted spatial discontinuities such as faults and fractures. The 3D view of the final migration results delineates an exquisite subsurface shallow structure with seafloor topography. In the vertical sections and time slices, a spatially continuous fault distribution is clearly observed.

5. Conventional 2D seismic and the latest 3D-HRS data were jointly used to decipher the subsurface structure and sedimentary facies in the Beppu Bay basin. On the 2D sections, a series of high-angle faults having a flower-like appearance extend near the southern bay coast and are interpreted as active traces of the laterally moving MTL. A releasing bend of the MTL forms a pull-apart sag around the bay bottom accompanied by numerous normal fractures at shallow depths, whereas a compressional bulge is emerging on the southern side of the bay, suggesting complicated and transient stress states during the late Quaternary. Such a tectonic context brought about westward migration of depocenters within the bay. Based on the general development history of the HVZ, the lower, middle, and upper seismic horizons are assigned to 5–6 Ma (initial stage of the HVZ), 0.7, and 0.3 Ma, respectively. We identified three auxiliary reflectors in the upper part of the sediment pile and correlated them with oxygen isotope stages 15–11. On such a stratigraphic basis, paleoenvironments in the latest Pleistocene are discerned by means of an amplitude RMS attribute analysis of 3D-HRS data. This method successfully delineated a river channel buried during the last glacial period, which is probably connected with the present Onogawa River. Structural interpretation of the 3D-HRS data indicates that a part of the ancient channel was cut by superficial fractures that are still growing.

It is noted that Beppu Bay is not an ideal place to execute such a comprehensive study. It has not been studied with deep boreholes, which can provide a firm stratigraphic basis for its burial history. Also, frequent influx of volcaniclastic debris degrades the seismic data quality. Vigorous activity on many faults obscures the detailed structural features due to spatiotemporal variance in the stress-strain state. We, however, succeeded in presenting a real image of tectonic basin development in this book based on integrated research. We hope our work stands as a precedent for the further challenge of understanding mobile belt frontiers, where intensive movements hinder efforts to decipher the Earth's evolutionary processes.

Author details

Yasuto Itoh
Osaka Metropolitan University, Osaka, Japan

*Address all correspondence to: yasutokov@yahoo.co.jp